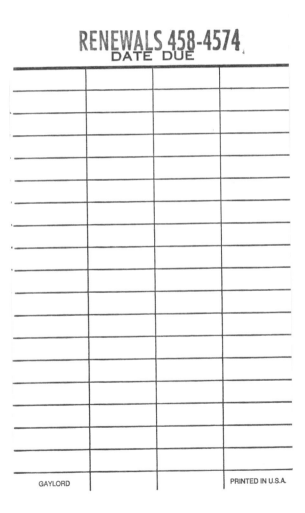

BIOLOGICAL SYSTEMATICS

BIOLOGICAL SYSTEMATICS

PRINCIPLES AND APPLICATIONS

Randall T. Schuh

George Willett Curator of Entomology and Chair
Division of Invertebrate Zoology, American Museum of Natural History
New York, New York
Adjunct Professor, Department of Entomology, Cornell University, Ithaca
and Department of Biology, City College, City University of New York

Comstock Publishing Associates a division of

Cornell University Press | Ithaca and London

First published 2000 by Cornell University Press

Printed in the United States of America

Cornell University Press strives to use environmentally responsible suppliers and materials to the fullest extent possible in the publishing of its books. Such materials include vegetable-based, low-VOC inks, and acid-free papers that are recycled, totally chlorine-free, or partly composed of nonwood fibers. Books that bear the logo of the FSC (Forest Stewardship Council) use paper taken from forests that have been inspected and certified as meeting the highest standards for environmental and social responsibility. For further information, visit our website at www.cornellpress.cornell.edu.

Library of Congress Cataloging-in-Publication Data

Schuh, Randall T.
Biological systematics : principles and applications / Randall T. Schuh.
p.   cm.
Includes bibliographical references (p.   ) and index.
ISBN 0-8014-3675-3 (cloth)
1. Biology Classification.   I. Title.
QH83.S345   2000
570′.1′2 — dc21   99-42377

Cloth printing   10   9   8   7   6   5   4   3   2   1

CONTENTS

Preface   vii

Acknowledgments   ix

Section I: BACKGROUND FOR THE STUDY OF SYSTEMATICS

Chapter 1 Introduction to Systematics   3

Chapter 2 Nomenclature   29

Chapter 3 Systematics and the Philosophy of Science   44

Section II: CLADISTIC METHODS

Chapter 4 Homology and Rooting   63

Chapter 5 Character Analysis and Selection of Taxa   89

Chapter 6 Quantitative Cladistic Methods   111

Chapter 7 Evaluating Results   146

Section III: APPLICATION OF CLADISTIC RESULTS

Chapter 8 Formal Classifications and Systematic Databases   165

Chapter 9 Historical Biogeography and Host–Parasite Co-evolution   179

Chapter 10 Ecology, Adaptation, and Evolutionary Scenarios   199

Chapter 11 Biodiversity and Conservation   209

Glossary   217

Subject Index   227

Author Index   234

Selecting and Acquiring Software

PREFACE

All fields of science have undergone revolutions, and systematics is no exception. For example, the discovery of DNA structure fundamentally altered our conception of the mechanisms of inheritance. One might assume that the most recent revolution in systematic biology would have come about through the proposal of a coherent theory of organic evolution as the basis for recovering information on the hierarchic relationships observed among organisms. Such was not the case, however, no matter the frequency of such claims. Rather, it was the realization by Willi Hennig — and others — nearly one hundred years after the publication of the *Origin of Species* by Charles Darwin, that homologies are transformed *and* nested and that phylogenetic relationships can best be discovered through the application of what have subsequently come to be called cladistic methods. The fact that the theory of evolution allowed for the explanation of a hierarchy of descent was seemingly not sufficient to arrive at a method for consistent recovery of genealogical relationships. It can further be argued that neither was it necessary.

The revolutionary changes did not stop there, however. At the same time that the methods of cladistics were changing taxonomic practice on how to recognize natural groupings, the issue of quantification was being discussed with equal fervor. Whereas systematics was long a discipline marked by its strong qualitative aspect, the analysis of phylogenetic relationships is now largely quantitative.

The introduction of quantitative methods to systematics began with the "numerical taxonomists." Their approach to grouping was based on overall similarity concepts, and the attendant assumption of equal rates of evolutionary change across phyletic lines. Establishment of systematic relationships is now dominated by cladistic methods, which use the parsimony criterion to form groups on the basis of special similarity and allow for unequal rates of evolutionary change. The logic and application of the quantitative cladistics were in large part developed by James S. Farris.

The overall approach of this book is to present a coherent and logically consistent view of systematic theory founded on cladistic methodology and the principle of parsimony. Some of its subject matter is in a style that would commonly be found in research papers, that is, argument and critique. This approach allows material to be presented in its unadulterated form rather than in the abstract, such that sources of ideas at which criticism is being directed are not obscured and can be found readily in the primary literature. The tradition of critical texts in biological systematics was established by Blackwelder, Crowson, Hennig, Sokal and Sneath, and others. I hope that the style of this book will help students see argumentation in science for what it is, a way of developing knowledge and understanding ideas. The alternative would be to obscure historical fact by pretending that the formulation of a coherent body of systematic thought has proceeded in a linear fashion, without sometimes acrimonious debate.

Organization of the Text. This work is divided into three sections, representing more or less logical divisions of the subject matter. Section I, Background for the Study of Systematics, comprises three chapters, which offer, respectively, an introduction to biological systematics, binomial nomenclature, and the philosophy of science as applied to systematics. Section II, Cladistic Methods, outlines the methods of phylogenetic analysis, with chapters on homology and rooting, character analysis, computer-implemented phylogenetic analysis, and evaluation of phylogenetic results. Section III, Application of Cladistic Results, comprises chapters on the preparation of formal classifications, historical biogeography and co-evolution, testing evolutionary scenarios, and biodiversity and conservation. A terminal glossary provides definitions for the specialized terminology of systematics used in this book.

Each chapter ends with lists of Literature Cited and Suggested Readings. The references cited in the text are those needed to validate an argument, but do not in all cases necessarily represent the most useful available sources. The Suggested Readings are intended to fill the gap between material presented and knowledge expected of a more sophisticated and inquiring reader. The readings are chosen for their breadth and quality of coverage, with consideration also being given to their accessibility. Most should be available in major university libraries, and thus be readily available to the majority of students and professors using this book.

Software

This book will be most successful as a teaching tool if used in conjunction with appropriate phylogenetic software. Only under such an approach can real-world phylogenetic problems be solved and understood during the process of studying cladistic methods. Software choices are described and the sources from which they can be purchased are listed at the back of the book.

ACKNOWLEDGMENTS

Several colleagues and friends provided discussion, assistance, advice, reviews, and encouragement during the writing of this book. For reviews of an early version of the manuscript, or parts thereof, I thank James Ashe, Gerasimos Cassis, Steven Keffer, David Lindberg, Norman Platnick, James Slater, Christian Thompson, Quentin Wheeler, and Ward Wheeler. For detailed reviews of the complete manuscript, I offer special thanks to Andrew Brower, James Carpenter, Eugene Gaffney, Pablo Goloboff, Dennis Stevenson, and John Wenzel. Dennis Stevenson gave me much advice on botanical examples and nomenclature and offered timely encouragement as the project progressed. My conception of issues of philosophy and systematic theory, as presented in this volume, has been influenced by discussions with Ronald Brady, Andrew Brower, James Carpenter, Eugene Gaffney, Pablo Goloboff, and Norman Platnick. Pablo Goloboff aided in clarifying my presentation of the quantification of cladistics and provided much of the material used in Chapter 6. Gregory Edgecomb offered suggestions on relevant literature. The students and auditors in my spring 1998 Principles of Systematics course at the City University of New York tested a version of the manuscript. Pamela Beresford, Christine Johnson, and Pablo Goloboff read and commented on the submitted version of the manuscript; Christine prepared the figures. Whatever the inputs from others, in the end, I am solely responsible for the final form of all arguments presented in the text.

The development of my views on the nature of systematics was shaped by two people in particular, my long-time friends and colleagues James S. Farris and Gareth Nelson. Since 1967, they, more than any other individuals, have profoundly affected our understanding of systematic theory. Thus, in an indirect way, they have greatly influenced the way I have written this book.

The encouragement of Steven Keffer and of my wife, Brenda Massie, caused me to undertake this project. Their confidence that I could produce a useful final product spurred me on. My young daughter, Ella, has been a patient helper during the preparation of the manuscript. The term 'systematics' is now indelibly imprinted in her mind.

RANDALL T. SCHUH

BACKGROUND FOR THE
STUDY OF SYSTEMATICS

1

Introduction to Systematics

Historical Setting

Systematics is the science of biological classification. It embodies the study of organic diversity and provides the tools to study the historical aspects of evolution. In this chapter we will explore the nature of systematics as an independent discipline and briefly survey the literature sources most frequently used by systematists.

The ancient Greeks produced, beginning in about 400 B.C., the first writings in the Western world that might be classed as scientific by modern standards. Many of the contributions of Plato, his student Aristotle, and others were translated into Latin by the Romans, and also into Arabic, whereby they received wider distribution, and by which mechanism many of them survived to modern times. Even though these important writings had great influence in their day, they remained obscure for about ten centuries until being "rediscovered" in the Middle Ages. It was the rediscovery, of Aristotle's work in particular, that rekindled interest in the thought processes that led to the development of modern science.

The exact nature of Aristotle's contribution to the field of systematics is a subject of varied interpretation, some of it positive, some of it negative, as we will see later on. What is not in dispute is the fact that Aristotle made some of the most detailed observations of the living world during his time, particularly with regard to animals.

Systematics — what is often called *taxonomy* — as currently practiced has its beginning in the work of the Swedish botanist and naturalist Carolus Linnaeus (Carl von Linné), and his contemporaries, in the mid-eighteenth century. Linnaeus' work built on the earlier contributions of authors such as the sixteenth-century Italian physician Andrea Caesalpino and the mid-seventeenth-century English naturalist John Ray.

The detailed history of systematics is a subject in its own right and would shed much light on how current systematic knowledge and the methods used to acquire that knowledge achieved their current form. However, much of that history is beyond the intended scope of this book. References dealing with the subject are included at the end of the chapter under "Suggested Readings." For our purposes, most of the history essential for understanding the current state of affairs in systematics dates from about 1950. Therefore, it is the relatively modern literature we will examine in detail.

The essential activities of systematics are of three basic types and have changed little over the last 250 years. *First,* is the recognition of basic units in nature, what are usually called *species.* Our understanding of the perpetuation of species has advanced greatly since the time of Linnaeus, primarily because of improved knowledge of the mechanisms of inheritance. Yet, with more than 2 million species of plants and animals, it has not been possible to study all of them in detail. Consequently, species are often recognized on the basis of morphological and other characteristics observable in specimens much as they were in the time of Linnaeus. The details of how species are recognized in mammalogy, entomology, bryology, and other fields of specialization are beyond the scope of this book.

Second, is the classification of those species in a hierarchic scheme. The natural hierarchy has long been recognized, and formal hierarchic classifications of plants and animals as published by Linnaeus existed well before the widely accepted formalization of a theory of organic evolution as first developed by Charles Darwin and Alfred Russell Wallace in the mid-nineteenth century. The authors of early classifications, as for example Linnaeus, were less highly motivated to investigate the forces that produced organic diversity than they were to describe the products themselves. What changed with the introduction of the Darwinian theory of organic evolution was that the observed relationships among organisms were then usually explained as being the results of the evolutionary process rather than divine creation. This change did not contribute, however, to producing a well articulated set of methods for deducing the relationships that most investigators began to assume were phylogenetic. That development had to wait nearly 100 years, and is the subject matter of much of this book.

Third, is the placement of information about species and their classification in some broader context, a subject to which we will return in the final chapters of this book.

The Schools of Taxonomy

The Evolutionary Taxonomic Point of View

As of the early 1950s, taxonomic theory was heavily influenced by what was known as the "new systematics," from which perspective the study of populations and infraspecific variability represented the crucial element for understanding biological diversity. The "new systematic" approach can be appreciated by examining the 1953 textbook *Methods and Principles of Systematic Zoology* by ornithologist Ernst Mayr and entomologists Gorton Linsley and Robert Usinger (1953). In that work, discrimination of species and subspecies occupied most of the substantive discussion of methods. About one-third of the volume dealt with nomenclature. Approximately two pages were devoted to the connection between characters and classification, and virtually no space was allocated to the discussion of techniques for discovering relationships among groups of organisms, be they species or groupings above the species level. The book contained only eight illustrations intended to portray phylogenetic relationships. In the largely neontological perspective of Mayr, Linsley, and Usinger, knowledge of relationships among organisms was heavily tied to a microevolutionary view of organic diversification and was presumably thought to flow directly from it.

A more strongly paleontological view of systematic biology, but one nonetheless closely associated with "evolutionary taxonomy," was portrayed a few years later in George Gaylord Simpson's *Principles of Animal Taxonomy* (1961). Simpson, a specialist on fossil mammals, devoted substantial space to the discussion of interrelationships among groups of organisms, while also emphasizing the importance of the temporal perspective that could be gained from geology and the study of fossils. He observed that "The construction of formal classifications of particular groups is an essential part and a useful outcome of the taxonomic effort but is not the whole or even the focal aim. The aim of taxonomy is to understand the grouping and interrelationships of organisms in biological terms" (p. 66). Simpson's perspective was that "Taxonomy is a science, but its application to classification involves a great deal of human contrivance and ingenuity, in short, of art."

An expanded understanding of the evolutionary taxonomists' perspective can be gained by examining passages from *The Growth of Biological Thought* (Mayr, 1982:209), where the author noted:

> That Darwin was the founder of the whole field of evolutionary taxonomy is realized by few . . . the theory of common descent accounts automatically for most of the degrees of similarity among organisms . . . but also . . . Darwin developed a well thought out theory with a detailed statement of methods and difficulties. The entire thirteenth chapter of the *Origin* is devoted by him to the development of his theory of classification.

A few pages later (p. 213), in what would appear to be a direct contradiction, Mayr stated, "As far as the methodology of classification is concerned, the Darwinian revolution had only minor impact." Mayr summarized his opinions by noting that Darwin's decisive contributions to taxonomy were that the theory of common descent provided an explanatory theory for the Linnaean hierarchy and that it bolstered the concept of continuity among organisms.

Classical evolutionary taxonomy, the approach advocated by Mayr and Simpson, argues for the portrayal of the maximum amount of *evolutionary* information in biological classifications. In the words of another proponent of this approach, "formal classification is an attempt to maximize simultaneously the two semi-independent variables of genetic similarity and phylogenetic sequence," with the caveat that a one-to-one correspondence between classification and phylogeny is impossible (Bock, 1974:391). Bock further opined that improvements in comprehension of systematic relationships among organisms should come through the more thorough study of organismal attributes, not through the introduction of new philosophical approaches. Not everyone was in agreement with Bock, and two opposing philosophical approaches were being argued in the literature at the time of his writing.

The Phenetic Point of View

Publications by a group of "extreme empiricists" propounding a taxonomy based on the concept of overall similarity began to appear in the late 1950s. The first textbook style summary of this body of thought was entitled *Principles of Numerical Taxonomy* (Sokal and Sneath, 1963). Although this approach was initially called "numerical taxonomy," the term *phenetic* — as introduced by Mayr in 1965 — seemed preferable because quantitative taxonomic approaches were not restricted to the phenetic school of thought. The pheneticists — including R. R. Sokal, P. H. A. Sneath, A. J. Cain, P. J. Harrison, F. J. Rohlf, D. H. Colless and others — were motivated to make taxonomy objective and "operational," the ultimate goal being to produce general purpose classifications in which relationships among groups of organisms are formed on the basis of overall similarity.

Hull (1970) summarized the views of the pheneticists as including (1) the desire to completely exclude evolutionary considerations from taxonomy because in the vast majority of cases phylogenies are unknown; (2) the belief that the methods of the evolutionists were not sufficiently explicit and quantitative; and (3) the observation that classifications based on phylogeny are by their very nature designed for a special purpose.

The methods of phenetics are numerous and have never been consistently codified, but are supposed to be atheoretical. Pheneticists have been referred to as neo-Adansonians because their methods are said to be patterned after the views of the eighteenth-century French botanist Michel Adanson.

The phenetic approach, as expounded by Robert Sokal and Peter Sneath (1963), was based on the precept that classifications incorporating the maximum number of unweighted observations would be general purpose, rather than being disposed toward some particular scientific theory, such as that of organic evolution. "Operationalizing" taxonomy would, in the view of the pheneticists, make the process of data gathering unbiased and possibly amenable to automation and the use of computers.

Phenetic techniques are implemented by converting the numbers of differences in attributes into "distances" which serve as a way to measure similarity among the groups of organisms under study. The distances are computed by counting the number of character differences between all possible pairings of taxa.

The Phylogenetic (Cladistic) Point of View

The German entomologist and systematic theorist Willi Hennig believed that "The task of systematics is the creation of a general reference system and the investigation of the relations that extend from it to all other possible and necessary systems in biology" (Hennig, 1966:7). He first propounded a set of 'methods and principles' in a 1950 work entitled *Grundzüge einer Theorie der phylogenetischen Systematik,* which was later modified and published in English under the title *Phylogenetic Systematics* (Hennig, 1966). In distinct contrast to the pheneticists, Hennig argued for an approach that he believed would directly reflect information concerning the results of the evolutionary process. Hennig viewed the hierarchic classifications long produced by systematists as the general reference system of biology, but he argued that the utility of those classifications could only be maximized if they accurately reflected the phylogenetic relationships of the organisms involved. This last point was not accepted by the pheneticists, nor was it accepted by the evolutionary taxonomists, as seen in Simpson's belief that "classification involves a great deal of human contrivance and ingenuity, in short, of art." Hennig's approach, at first labeled *phylogenetic systematics —* but today called *cladistics —* forcefully articulated the idea that genealogical relationships should be based on special similarity (shared derived characters) and that those relationships should be faithfully reflected in a formal hierarchic listing.

In fairness to history, the methods described by Hennig had apparently been applied earlier — as for example by P. C. Mitchell working with birds (1901; as cited in Nelson and Platnick, 1981) and by W. Zimmermann (1943) working on plants. And, as was pointed out by Platnick and Cameron (1977), the fields of textual criticism (stemmatics) and historical linguistics both use methods nearly identical to those propounded by Hennig for establishing historical relationships among manuscripts and languages, respectively. Thus, the approach of grouping by special similarity seems to have a general applicability to systems involving lineage evolution and diversification over time. Within biology, the earlier applications of the cladistic approach — as by Mitchell — seem to have been insufficiently influential to revolutionize systematics.

Sidebar 1
Clades, Cladistics, Cladists, and Other Terms

The term *clade* was used in the 1950s by Huxley to denote an evolutionary lineage. It was adopted in the form *cladistics* and applied to phylogenetic systematic studies of the type espoused by Willi Hennig as early as 1965 by Camin and Sokal and also by Mayr. The graphical depictions of phylogenetic relationships produced by these methods were called *cladograms* by those same authors. The term *cladist* was soon in use, initially often as a pejorative, to refer to those who used the methods of Hennig.

At the same time that 'cladistics' was coined, Mayr, Sokal, and Camin used the term *phenetics* for an approach previously widely known as numerical taxonomy. Mayr made it clear that it was the methods and their justification that were distinctive, not the use of numerical techniques. The diagrams of relationships produced with phenetic techniques were called *phenograms* by Mayr and Camin and Sokal. Those who practiced phenetics were soon called *pheneticists*.

The additional terms *syncretist* and *gradist* are also found in the literature. They usually refer to individuals whose approach to taxonomy reflects a combining of methodologies into what is often called *evolutionary taxonomy*.

How Do Evolutionary Taxonomy, Phenetics, and Cladistics Differ?

It may seem paradoxical that in 1965, after more than 200 years, the field of systematics still did not have a clearly codified — and broadly accepted — set of methods. Yet, that was indeed the case.

Let us pose three questions as a way of examining the basic precepts of the three "schools" introduced above, each of which is arguing for the primacy of its point of view as the most efficacious approach to the study of biological systematics.

1. *Can we reconstruct phylogeny?* The phenetic point of view was clearly "No," whereas evolutionary taxonomists and cladists felt just the opposite.
2. *Does evolutionary change proceed at the same rate in different lineages?* The pheneticists thought that it did, or at least they applied methods that assumed that it did. The evolutionary taxonomists thought that rates varied and wished to incorporate that information in their results. The cladists applied methods that were unaffected by variation in rates and also came to conclusions distinct from those of the other two schools about how the results of differences in rates of divergence across lineages might best be portrayed in formal classifications.

3. *Are all attributes of organisms useful in forming classifications?* Pheneticists thought the answer was "yes." Their methods explicitly used techniques that would measure the degree to which groups were similar *and* different, these measures being mediated by the assumption of constant rate of change. Cladists took the view that groupings could only be formed on the basis of unique attributes, for to do otherwise would be to allow any possible grouping. Many evolutionary taxonomists accepted Hennig's cladistic point of view concerning the formation of unique traits in certain lineages during the course of evolution, but maintained that degree of difference among lineages should be recognized in assigning rank in formal classifications.

The salient attributes of the three taxonomic approaches are characterized in Table 1.1. Figure 1.1 shows phenetic and cladistic approaches for presenting data and the results of analyzing those data. As can been seen, cladistic methods produce groupings determined only by the presence of attributes unique to the group being formed. To the contrary, phenetic methods form groupings on degree of difference, with the consequent discarding of information on attributes that are unique to groups. Thus, groups A + B and C + D are found in the cladogram, but neither grouping is shown in the phenogram. This is because B is so different from A, and D is so different from C, that neither forms a group with its "nearest relative," but rather A and C group together because they are less different from each other than either is from the other two taxa. Thus, under phenetic methods, a large number of unique attributes (e.g., characters 6–14 in taxon B, and characters 15–20 in taxon D) will cause groups to be formed (A + C), even though the members of such groups share no unique attributes in common. In contrast, cladistic methods only form groups on the basis of shared unique attributes (group A + B, character 1; group C + D, character 3) and treat attributes unique to a single taxon (as in taxa B and D) as irrelevant to the process of group formation. The evolutionary taxonomic groupings are the same as

Table 1.1. Attributes of the "three schools" of systematics

	Phenetics	Evolutionary taxonomy	Cladistics
Data type	Character data converted to matrix of distances between taxa	Discrete characters	Discrete characters
Grouping method	Overall similarity	Special similarity	Special similarity
Diagram type	Phenogram	Evolutionary tree	Cladogram
Hierarchic level determined by	Amount of difference	Amount of difference	Sharing of unique attributes
Sensitive to rate differences	Yes	Yes	No

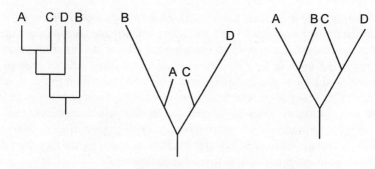

Phenogram, showing levels of clustering based on overall similarity.

Evolutionary tree, showing genealogical relationships and degree of divergence.

Cladogram, showing genealogical relationships solely as recency of common ancestry.

CHARACTER MATRIX

Character Taxon	1	2	3	4	5	6	7	8	9	10	11	12	13	14	15	16	17	18	19	20
Group X	0	0	0	0	0	0	0	0	0	0	0	0	0	0	0	0	0	0	0	0
Group A	1	1	0	0	0	0	0	0	0	0	0	0	0	0	0	0	0	0	0	0
Group B	1	0	0	0	0	1	1	1	1	1	1	1	1	1	0	0	0	0	0	0
Group C	0	0	1	1	1	0	0	0	0	0	0	0	0	0	0	0	0	0	0	0
Group D	0	0	1	0	0	0	0	0	0	0	0	0	0	0	1	1	1	1	1	1

DISTANCE MATRIX

	A	B	C	D
A	0			
B	10	0		
C	5	13	0	
D	9	17	8	0

Fig. 1.1. Cladistic character matrix and the corresponding phenetic distance matrix for four groups of organisms. Group X in the character matrix determines the base of the evolutionary tree and cladogram. The distances are computed by counting the number of character differences between all possible pairings of taxa. The phenogram, evolutionary tree, and cladogram depicting relationships among the four groups are resolved through the methods of phenetics, evolutionary taxonomy, and cladistics, respectively. Note that in the diagrams, lineage relationships retained in the evolutionary tree and the cladogram are modified in the phenogram, as a result of the sensitivity of phenetic methods to variability in evolutionary rates in the different groups of organisms (modified from Farris, 1971).

the cladistic ones, but the degree of divergence (difference) is indicated by the relative lengths of the branches in the former.

Are There Really Three Schools?

The pheneticists argued for grouping on the basis of overall similarity, an approach implemented through the acquisition and analysis of as much "objective" data as possible. Their anti-phylogenetic stance and quantitative persuasion set them apart for a while, but it was soon pointed out that phenetics and classical evolutionary taxonomy actually shared a critical element in common — both

Sidebar 2
The Writings of Willi Hennig: From Relative Obscurity to Preeminence

The original works of Willi Hennig on 'phylogenetic systematics,' and his published examples applying that method, were not widely read and appreciated because they were written in German and appeared in a relatively obscure published form. Two 1966 publications, in English, changed this situation: Hennig's *Phylogenetic Systematics,* which was translated from a German-language manuscript, and a monograph by the Swedish entomologist Lars Brundin (1966) entitled *Transantarctic Relationships and Their Significance, as Evidenced by Chironomid Midges.*

The work of Brundin was unique in arguing persuasively for the merits of Hennig's methods, while simultaneously offering an application of them to a relatively complex real-world problem. The end result was a magnificent taxonomic and biogeographic example.

Gareth Nelson, late a postdoc from the Swedish Natural History Museum, personally transported the ideas of Hennig and Brundin to the offices and lecture halls of the American Museum of Natural History when he joined the staff of that institution as a curator of ichthyology in 1967. He further invigorated discussion of the subject through his editorship of *Systematic Zoology* from 1973 through 1976. Whereas some American entomologists, such as Pedro Wygodzinsky, had independently become familiar with Hennig's approach, it was primarily Nelson's influence that changed the thinking of vertebrate systematists.

The transmission of phylogenetic methods to botanists has taken place more recently. As late as 1978, Bremer and Wantorp lamented that botanists had still not recognized the importance of Hennig's work and pointed to some of the glaring cases of paraphyletic groups, such as Dicotyeledoneae, still recognized in all major classifications of higher plants at that time.

approaches emphasized the importance of "overall similarity" in establishing rank in formal classifications. Cladistics, on the other hand, emphasized groupings based on "special similarity," what Hennig called *apomorphy*. If the classical evolutionary taxonomists and cladists could have agreed on the properties of formal classifications, they might have agreed on the choice of methods. Such was not the case, however.

Therefore, it is the pheneticists and classical evolutionary taxonomists that must be associated with one another. If this association at first seems contradictory — having observed the biological gulf separating these two groups of practitioners — consider the following statements by Mayr (1982:230): Cladistic "classifications are based entirely on synapomorphies [shared special similarities], even in cases, like the evolution of birds from reptiles, where the autapomorph characters [unique special similarities] vastly outnumber the synapomorphies with their nearest reptilian relatives," and (p. 233) "the main difference between [classical evolutionary taxonomy] and cladistics is in the considerable weight given to autapomorph characters [by the former]." In other words, because birds have so many attributes unique to themselves, they should not be classified as a subgroup of Reptilia even though their novel features are simply modifications of the more general attributes found in crocodiles, lizards, and other 'reptiles.' Mayr, in his role as one of the most vocal proponents of classical evolutionary taxonomy, made it clear in this statement that unique characters must be considered in the ranking of groups in classifications. Mayr cited examples to demonstrate why excluding unique characters in assessing rank produces what he believed to be absurd results. These examples included the grouping of man with the great apes and, more poignantly for him as an ornithologist, the treatment of birds and crocodiles as each others' nearest relatives among living organisms, or of birds as a subgroup of dinosaurs when extinct groups are also considered. The apparent dismissal of similarity information with incorporation of birds into the Reptilia through the use of cladistic methods is what Mayr argued against, and it is exactly that methodological attribute that distinguishes cladistics from classical evolutionary taxonomy and phenetics.

To further illustrate these points, consider the following classifications of the living amniote tetrapods as they might be produced by the three schools of taxonomy:

Evolutionary taxonomic classification:
    Class Mammalia
    Class Reptilia
        Subclass Testudines (turtles)
        Subclass Squamata (lizards, snakes, crocodiles)
    Class Aves

Phenetic classification:
  Class Mammalia
  Class Reptilia
    Subclass Testudines
    Subclass Squamata
  Class Aves

Cladistic classification:
  Class Mammalia
  Class Reptilia
    Subclass Testudines
    Subclass Sauria
      Infraclass Squamata (lizards and snakes)
      Infraclass Archosauria (crocodiles, birds)
        Order Crocodilia
        Order Aves

Even though the evolutionary taxonomists are aware that birds are phylogenetically part of the Reptilia, their desire to express degree of difference in formal classifications obscures that relationship. Only the cladistic classification correctly portrays genealogical connections — groupings based on the shared possession of distinctive attributes — in the formal classification.

Although our description of classificatory methods has so far been based largely on the work of zoologists, this is only because it is in their works that we find the clearest enunciation of classificatory methods. In botany there is a tremendous amount of comprehensive work, but the guiding force in its preparation seems to be based on tradition rather than a more analytic approach to the justification of methods. Nonetheless, the nature of the methods used in many botanical classifications is clear, as it is in zoology. Consider for example the classification of seed plants at the highest levels.

Traditional classification (e.g., Lawrence, 1951):
Phylum Spermatophyta
  Subphylum Gymnospermae
    Class Cycadales
    Class Ginkgoales
    Class Coniferales
    Class Gnetales
  Subphylum Angiopermae
    Class Dicotyledoneae
    Class Monocotyledoneae

Cladistic classification (Loconte and Stevenson, 1990, 1991):
Phylum Spermatophyta
  Subphylum Cycadales
  Subphylum Cladospermae
    Infraphylum Ginkgoales
    Infraphylum Mesospermae
      Microphylum Coniferales
      Microphylum Anaspermae
        Class Gnetales
        Class Angiospermae
          Subclass Calycanthales
          Subclass [Unnamed]
            Infraclass Magnoliales
            Infraclass [Unnamed]
              Microclass Laurales
              Microclass [Unnamed]
                etc.

The Gymnospermae in the traditional classification are members of a group without any characteristics distinctive to them, but rather are recognized by having seeds with two cotyledons and not having flowers. Obviously all Spermatophyta have seeds and most have two cotyledons; therefore these attributes are not distinctive for the Gymnospermae. The Dicotyledoneae, the classification of which is listed only in part, are defined in a similar way: plants with flowers and two cotyledons. But all seed plants other than monocots have two cotyledons; therefore the characteristics of the Dicotyledoneae are more accurately stated as plants with flowers and without seeds with one cotyledon. Cladistic methods have shown that these, and other long-recognized higher groupings within the green plants, are not defined on the basis of distinctive attributes, but rather on a combination of attributes, some of which also occur in other groups. Clearly, the methods, whether explicit or not, are phenetic.

As currently understood, then, approaches to biological classification can be recognized as being of two obvious types: those that treat overall similarity as important in group recognition and those that group by special similarity alone. Alternatively stated, the former approach adheres to the idea that the amount of apparent difference among groups is important in reflecting relationships, whereas the latter does not. It is the cladistic approach that now dominates systematic thinking and forms the major focus of this book. We will explore further the reasons for this dominance in Chapter 3, as well as delving into the last strongholds of grouping by overall similarity when discussing the analysis of molecular data.

Some Terms and Concepts

In order to continue effectively our discussion of systematics, it will be help-ful to clarify the meanings of several commonly used terms and to make further observations on the place of systematics within biology.

The rigid definition of terms is usually less important than understanding the context in which they are used; nonetheless, it is important to understand the meanings to be ascribed to certain terms in this book. Such an approach will help us to better comprehend the types of scientific problems systematists attempt to solve as well as to gain additional perspective on scientific contributions of sys-tematics as a field.

Taxonomy and *systematics* are terms that embody the activities of systema-tists, but the exact meanings ascribed to them have varied widely. Referring once again to the work of Simpson (1961: 7), we find the following definitions:

> Taxonomy is the theoretical study of classification, including its bases, principles, procedures, and rules.

> Systematics is the scientific study of the kinds and diversity of organisms and of any and all relationships among them.

As Simpson's definitions suggest, systematics has often been used as the more inclusive term, and taxonomy and classification would therefore be subsumed within it. Some have argued that systematics is the term having historical prece-dence and that taxonomy should therefore be supplanted. Politically artful use of the terms has at times suggested that taxonomy somehow represents a mun-dane activity most closely associated with identification and that systematics is the more elevated form of the discipline. Among such usages could be included "biosystematics," a term implying the integration of a broader range of biologi-cal information than would be the case within "ordinary" systematic studies and a consequent elevation of the status of the field (e.g., Ross, 1974).

Systematics is the term used most frequently in this work, because of its wide usage and broad connotation. *Taxonomy* would, nonetheless, serve equally well. The field of study subsumed under these terms encompasses the methods and practices of describing, naming, and classifying biological diversity, at the spe-cies level and above.

Classification represents the codification of the results of systematic studies. According to Simpson (1961:9), ". . . classification is the ordering of animals into groups (or sets) on the basis of their relationships, that is, of associations by con-tiguity, similarity, or both." The issue of contiguity was resolved, at least in large part, with the proposal by Charles Darwin and Alfred Russell Wallace of an ex-plicit and widely accepted theory of organic evolution. The role of similarity has

KEY TO MALES OF NEARCTIC
ATRACTOTOMUS

1. Hind femora with moderate to dense covering of appressed, scalelike setae (figs. 81, 82), or rarely with scalelike setae restricted to ventral surface of femora (fig. 91) 2
    Hind femora without scalelike setae . . 10

2(1). Hemelytral membrane with scalelike setae, usually most abundant inside areolar cells and along veins (fig. 13) . . 3
    Hemelytral membrane without scalelike setae 4

3(2). Antennal fossae nearly contiguous with anteroventral margin of eye; length of antennal segment II slightly greater than width of head across eyes; peritremal disk and coxae pale . . . .
    . . . . *taxcoensis*, new species
    Antennal fossae removed from anteroventral of eye by distance equal to or greater than diameter of antennal segment I (fig. 7); length of antennal segment II much less than width of head across eyes (ratio—0.46:1 to 0.62:1); peritremal disk and coxae dark, or disk rarely somewhat paler than adjacent thoracic sclerites . . . . *balli* Knight

4(2). Hind femora with scalelike setae restricted to narrow band on ventral surface (fig. 91) . . . . *tuthilli* (Knight)
    Hind femora with more or less generally distributed scalelike setae (figs. 81, 82) 5

5(4). Tibiae uniformly dark reddish brown or black, never paler than adjoining femora; antennal segment III uniformly darkened, without pale region basally; dorsum uniformly dark brown or black, without red or yellow markings . . . 6
    Tibiae, at least distally, yellow or brownish yellow, rarely somewhat darker, but always paler than adjoining femora; antennal segment III uniformly pale yellow to yellowish brown, or with distinct pale region basally; dorsal coloration variable, usually with at least bases of corium and clavus, embolium, and cuneus yellowish brown or red 8

6(5). Antennal segment II strongly inflated, greatest thickness nearly twice that of segment I (fig. 41); length of gonopore sclerite in lateral view approximately 1.5 times that of the gonopore (figs. 140, 141) . . . . *reuteri* Knight
    Antennal segment II linear or weakly clavate, not strongly inflated, greatest thickness rarely little more than that of segment I (figs. 18, 22); length of gonopore sclerite in lateral view approximately twice that of the gonopore (figs. 120, 121, 123, 124) 7

7(6). Ratio of length of antennal segment II to width of head across eyes from 0.85:1 to 0.92:1; vesica as in figures 120 and 121, spinose field on gonopore sclerite usually broad proximally . . . .
    . . . . *arizonae* (Knight)
    Ratio of length of antennal segment II to width of head across eyes from 0.73:1 to 0.80:1; vesica as in figures 123 and 124, spinose field on gonopore sclerite usually narrow proximally . . . .
    . . . . *cercocarpi* Knight

8(5). Ratio of length of antennal segment II to width of head across eyes from 0.86:1 to 0.90:1; vesica as in figure 160, with short gonopore sclerite, and gonopore well removed from apex of vesical strap . . . . *ramentum*, new species
    Ratio of length of antennal segment II to width of head across eyes from 0.72:1 to 0.83:1; vesica either with long gonopore sclerite (fig. 162), or gonopore located near apex of strap (fig. 122) . . 9

9(8). Vesical strap distad of medial coil elongate, gonopore removed from apex, gonopore sclerite with elongate row of evenly distributed spines (fig. 162) . . .
    . . . . *rubidus* (Uhler)
    Vesical strap distad of medial coil short, gonopore near apex, gonopore sclerite with spines mostly restricted to distal half (fig. 122) . . . . . . *atricolor* (Knight)

10(1). Hemelytral membrane with widely distributed scalelike setae (fig. 12) . . . . . .
    . . . . *acaciae* Knight
    Hemelytral membrane without scalelike setae 11

11(10). Dorsum without scalelike setae 12
    Dorsum with scalelike setae, sometimes restricted to anterior margin of pronotal disk and bases of clavus and corium 13

12(11). Head, pronotum, and base of hemelytra yellowish orange, sometimes tinged with red; remainder of hemelytra shiny black; ratio of width of vertex to width of head across eyes from 0.48:1 to 0.51:1 . . . .
    . . . . *chiapas*, new species
    Head, pronotum, and hemelytra uniformly reddish brown; ratio of width of vertex to width of head across eyes from 0.36:1 to 0.40:1 . . . .
    . . . . *polymorphae*, new species

13(11). Hemelytra with light and dark scalelike setae . . . . *nicholi* Knight
    Hemelytra with silvery white scalelike setae only 14

14(13). Antennal segment II yellow or brownish yellow, rarely with apex narrowly darkened 15

Fig. 1.2.

43. GALACTIA Adans. Fam. Pl. **2:** 322. 1763.

Scandent or erect herbs or shrubs; leaves pinnately 3 or 5-foliolate, the leaflets large or small; flowers small or large, usually racemose; fruit linear, bivalvate.

Leaflets 4 to 9 cm. wide. Plants scandent_____1. **G. viridiflora.**
Leaflets less than 3.5 cm. wide.
    Flowers in axillary clusters_____**2. G. brachystachya.**
    Flowers racemose.
        Racemes stout, dense, sessile, mostly shorter than the leaves.
                3. G. multiflora.
        Racemes slender, interrupted, pedunculate, mostly longer than the leaves.
            Plants erect; leaflets acute or acuminate_____**4. G. incana.**
            Plants scandent or trailing; leaflets often obtuse.
                Leaflets glabrous on the upper surface_____5. **G. acapulcensis.**
                Leaflets variously pubescent on the upper surface.
                    Leaflets bright green on the upper surface, not closely sericeous on
                      either surface_____**6. G. striata.**
                    Leaflets grayish, closely sericeous on both surfaces.
                      Leaflets white beneath with a soft silky pubescence, oval or ovate.
                              7. G. argentea.
                    Leaflets grayish beneath with rather stiff pubescence, usually
                      oblong_____**8. G. wrightii.**

Fig. 1.2. Examples of dichotomous keys. Groups are progressively subdivided using easily recognized characteristics, but those characteristics may have little value in forming natural groups. The indented format is frequently used in botanical works. Partial key of the insect genus *Atractotomus* from Stonedahl (1990; courtesy of The American Museum of Natural History). Key to the Mexican species of plant genus *Galactia* from Standley (1922).

been frequently invoked, but historically has had a much less well-clarified role in the construction of classifications. This issue is the subject of much of this book.

Within biology, the term classification usually refers to a "natural hierarchy," one which ostensibly reflects genealogical relationships among organisms. That is not to say that other types of classifications are not used by biologists or by other scientists. For example, special purpose classifications are commonly found in the form of keys (Fig. 1.2)—schemes usually designed to facilitate the identification of organisms. Such classifications are often based solely on the concept of similarity and may bear little resemblance to a "natural hierarchy."

Classifications recognizing natural classes are possibly best typified by the periodic table of the elements. This is a construct of great utility, but one which does not directly convey historical or hierarchic information.

As we will see, the activity of preparing formal classifications depends on the adoption of a set of procedures, as discussed in Chapter 8, rather than the direct resolution of a scientific problem. Nonetheless, biological classifications have great scientific value, although the limits of their utility, and the means by which they store and transmit information, have often been misunderstood. These misunderstandings have led to many erroneous criticisms concerning the flaws of taxonomy. We will discuss formal classifications further in Chapters 2, 3, and 8.

In sum, the question of whether a consensus of opinion exists as to the precise meanings of the terms *systematics, taxonomy,* and *classification* might best be answered by pondering the titles of influential works describing the methods and procedures of this general field of study. Among many possible examples are Simpson's *Principles of Animal Taxonomy,* Ross's *Biological Systematics,* Mayr's *Principles of Systematic Zoology,* and Crowson's *Classification and Biology.* These authors discussed essentially the same subject, but had little hesitation in using these three closely related, but different, terms to describe that subject.

Identification usually means "to place a name on." We might say that a specimen was identified as such-and-such species or that a specimen was "determined" to be such-and-such a species. Identification is an important day-to-day activity for most systematists, and possibly for most human beings, but the activity does not form the basis for recognition of groups or for establishing relationships among them.

The Place of Systematics within Biology

There is a frequently held perception on the part of the general public, and some scientists, that equates science with experiment. Much of science is experimental in nature, but this is not a prerequisite for qualifying as "scientific." Systematists make discoveries involving the natural world, but those discoveries are usually not the result of experiment but rather of observation and comparison. That science need not be experimental might be appreciated by observing that sequencing the human genome and making celestial observations with the Hubbell Space Telescope do not involve experiment, but nonetheless represent important and valid endeavors in science.

The field of biology has been divided in many ways. Not all of these divisions are easily compared, and indeed some are incompatible. The division can be organism based, such as botany and zoology. Or it might be practical, such as agriculture and medicine. These approaches have not served well to incorporate systematics as a field of study. More constructively we might divide academic biology into evolutionary and non-evolutionary, comparative and general, or reductionist and integrative.

Systematics is the most strongly comparative of all of the biological sciences, transcending the differences between botany and zoology. It is also the most strongly historical subdiscipline within biology and as such provides the basis for nearly all inferences concerning historical relationships. Among the earth sciences, systematics is directly comparable to historical geology, and indeed the two fields find integration in paleontology.

We might contrast systematics with "general biology," which often involves the study of a single organism or a single organ system. Such an approach is non-historical and places limited reliance on comparison among organisms and their organ systems.

The Units of Systematics

Taxa (singular, *taxon*) are the basic units of systematics. This term can be used to refer to a grouping of organisms at any level in the systematic hierarchy. It does not refer to individual organisms, although such are often studied by systematists as representatives of a given taxon. Because it is frequently necessary to refer to taxa without reference to their hierarchic position, a number of methods for making such reference have been developed. Thus, terms such as *terminal taxon* and *operational taxonomic unit* (OTU) have been proposed, the former being used primarily in cladistics, the latter arising out of phenetics. Although such terms may carry somewhat different connotations, depending on the author who is using them, the general sense is often the same. We might define a *terminal taxon* as *a group of organisms that for the purposes of a given study is assumed to be homogeneous with respect to other such groups.*

The biological literature is replete with discussions of speciation mechanisms and *species concepts.* For two reasons, these areas will not be discussed in detail in the current volume. First, the actual mechanistic explanations for species formation are irrelevant to most systematic conclusions. Second, although the search for a definition of *species* is a pervasive element in the biological literature, the formal codification of a species concept applicable across all of biology has proved elusive and maybe unnecessary. Systematists deal with recognizable or diagnosable taxa, be they at the minimum level (usually called species) or at some more inclusive level. Whatever they are called, if they are recognizably distinct, taxa at any level form valid units for systematic analyses. Many useful references on species and speciation exist, and the reader is encouraged to search them out (e.g., Cracraft, 1998; Ereshefsky, 1992; Futuyma, 1998). Furthermore, nearly any college course in population genetics or evolution deals with issues of species formation. Thus, readers of this volume should already be familiar with these general issues of evolutionary biology.

One might legitimately ask, however, if the approach of accepting some minimal-level taxon does not allow for the treatment of males and females as different taxa, or the treatment of different morphs of a polymorphic taxon as distinct taxa. The answer must be that such an approach would be naive, ignoring much of what biologists would hold to be self-evident — the contiguity of parents and their offspring, whether the parents are bisexual, asexual, or hermaphroditic. The practical aspects of studying the natural world at times confound this seemingly simple issue. For example, in certain groups of wasps, virtually independent classifications exist for the males and the females because current knowledge does not allow for unequivocal association of the two sexes of the same species. Species discrimination among clonal organisms or suspected hybrids can also be problematic. Indeed, the solutions adopted by specialists in one group of organisms may be quite different from those applied in another. In some cases the problem will be evident and therefore subject to resolution before detailed systematic studies commence; in other cases solutions to such problems will only be clarified through careful systematic study.

Taxa bear *characters,* and it is by these attributes that they are recognized (diagnosed; defined). It is the characters that are usually viewed as the data of systematics, what Ross (1974) referred to as the "material basis of systematics." *Species can be recognized by some unique combination of characters,* although many may also possess attributes unique to themselves. At the level of inclusive groupings, such as genera, or families, taxa cannot be diagnosed by this method because any and all possible combinations of groupings would thereby become possible. This produces untenable and relatively useless classifications. Thus, *inclusive groupings of two or more 'species' must be recognized on the basis of unique characters, which do not also occur in other groups.*

The practical problems of diagnosibility at the species level may take on a subjective quality. For example, earlier in this century, ornithologists recognized roughly 20,000 species of birds, whereas now they recognize about 9,000 species. Presumably, birds have neither evolved nor de-evolved during this short period of time. What has changed, however, is the concept of the minimal diagnosable unit. The number of bird species is beginning to rise again. This is not so much because new species are being discovered but because more critical studies have clarified that what were once thought to be variants within a 'species' should be interpreted as independent lineages in nature, lineages with distinct histories that can be studied with the tools of systematics (e.g., Cracraft, 1992).

The Systematic Literature

The systematic literature begins with the earliest works purporting to represent biological classifications. Because of the tremendous biological diversity dealt with by systematists, we are confronted with an almost overwhelming number of pertinent publications. Much of the systematic literature will not be found in local libraries, or even in many college or university libraries, but rather exists only in specialized research libraries. A brief survey of the types of publications containing information on biological systematics will help to orient a search for relevant books and articles.

Descriptive Works

Documentation of biological diversity has historically been a largely descriptive enterprise. This includes not only the recognition of plant and animal species and the higher taxa into which they are grouped but also the more detailed inquiry into the structural attributes (morphology, anatomy) of those organisms. The larger, more comprehensive publications of this type are often called *monographs* or *revisions* (Fig. 1.3). They have sometimes appeared as books, particularly before about 1840, but after that time more commonly as journal articles, the periodical literature of science.

Today there are thousands of journals that publish myriad articles on systematics. Many of these publications focus on a single group of organisms, others are restricted geographically, and still others may publish systematic papers among a broad spectrum of other scientific articles. Summaries of the descriptive literature often become available to the general public in the form of *handbooks, field guides, faunas, floras,* and similar sources that are written in a more easily understood fashion and present information in a more distilled manner.

Catalogs and Checklists

Comprehension of the literature dealing with animals and plants is a formidable task because of the sheer volume. This gulf is bridged by publications that are often referred to as *catalogs* (Fig. 1.4) and *checklists.* The exact form differs from group to group, but the general type of information contained in them is the same.

In botany, the historically most important source of this type is the *Index Kewensis,* which lists the names and place of publication of all seed-plant taxa at the family level and below. A new supplement is issued annually. The work was started by Joseph Dalton Hooker, then Director of the Royal Botanic Garden at Kew; the original four volumes were prepared with funds provided by Charles Darwin. The entire catalog, containing approximately 900,000 names, is now available on CD-ROM. Compact, readily available sources for the names of most plant genera are Mabberley (1997), Wielgorskaya (1995), and Willis (1973). *The Kew Record of Taxonomic Literature Relating to Vascular Plants* is an additional valuable source for locating botanical literature.

The situation in zoology is more complicated; no single source will serve for all groups, and indeed no definitive works exist for many groups. Nonetheless, one will find world or regional catalogs or checklists for many groups of animals. These works usually provide information on names, distributions, and, often, extensive lists of sources from which the information was derived. Examples include Eschmeyer (1990) for fishes, Sibley and Monroe (1990) for birds, and Wilson and Reeder (1993) for mammals. For many less well-known groups, it may be possible to find only a listing for Western Europe — if the groups occur there.

The type of structured information contained in catalogs lends itself to preparation in the form of computer databases. Although only a limited number of such databases are currently available, the situation will likely change dramatically within the next few years. Eventually treatments comparable to the

Mertila bhamo, new species
Figures 36, 38

DIAGNOSIS: Recognized by the broadly red-dened bases of clavus and corium; length of antennal segment II nearly equal to width of head across eyes; prominent tylus; and the structure of the male genitalia (fig. 38), especially the large, spinelike tubercles on basodorsal margin and inner-medial surface of right lateral process of genital capsule (fig. 38b, c).

DESCRIPTION: MALE. Length 6.25; dark, metallic coloration on distal portion of clavus and corium not extending anteriorly beyond level of apex of scutellum, embolium pale to near level of apex of corium. HEAD. Width across eyes 1.40, width of vertex 0.72; tylus prominent, strongly produced basally; length of antennal segment I 0.59, basal half light reddish brown, darkening to fuscous apically, segment II 1.35, dark reddish brown, very slightly expanded distally; labium damaged distally. PRONOTUM. Posterior width 1.90.

HEMELYTRA. Dark coloration on corium and clavus more brownish black. LEGS. Femora yellowish brown, tinged with red; tibiae brown; tarsi brown or brownish yellow. GENITALIA. Figure 38. FEMALE. Length 5.50; embolium less extensively pale than for male.

ETYMOLOGY: Named for the type locality; a noun in apposition.

DISTRIBUTION: Burma (fig. 36).

HOLOTYPE ♂: BURMA: **Bhamo District:** Bhamo, Aug. 1885, Fea (Distant collection, 1911-383) (BMNH). The above specimen was incorrectly identified as *malayensis* by W. L. Distant in the early 1900s. It also seems to be the specimen upon which Poppius (1912a) based a redescription of *malayensis*.

ADDITIONAL SPECIMEN: INDIA: 1♀, [port interception with no specific locality data], July 22, 1939, "on *Vanda*" (USNM). This specimen appears conspecific with the type (same dorsal coloration, prominent tylus, and second antennal segment), but a male example is needed for positive identification.

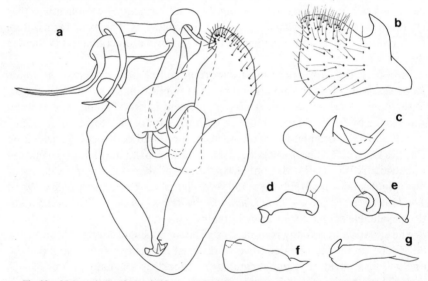

Fig. 38.   Male genitalia of *Mertila bhamo.* **a.** Genital capsule, posterior view. **b, c.** Right lateral lobe of genital capsule. b. Lateral view. c. Dorsal view. **d, e.** Left paramere. d. Lateral view. e. Posterior view. **f, g.** Right paramere. f. Lateral view. g. Dorsal view.

Fig. 1.3. Typical page from a revision, including diagnosis, description, holotype designation, locality data, and figures of structures important in characterizing a taxon, in this case described as new (from Stonedahl, 1988: 37, 38; courtesy of The American Museum of Natural History).

Gen. **Cyclosternum** Ausserer, 1871

♂♀ **kochi** (Ausserer, 1871)   Venezuela
   C. k. Schmidt, 1993d: 62, f. 55–56 (♂).

♀ **longipes** (Schiapelli & Gerschman, 1945)   Venezuela
   C. l. Schmidt, 1993d: 62, f. 51 (♀).

♀ **obesum** (Simon, 1892)   Brazil
   Magulla obesa Gerschman & Schiapelli, 1973b: 75, f. 14–15 (♀).
   C. o. Schmidt, 1993d: 62, f. 50 (♀).

♂ **rufohirtum** (Simon, 1889)   Venezuela
   Adranochelia rufohirta Gerschman & Schiapelli, 1973b: 62, f. 74–78 (♂).
   C. r. Schmidt, 1993d: 62, f. 57–60 (♂).

♂♀ **schmardae** Ausserer, 1871   Colombia, Ecuador
   C. s. Bücherl, Timotheo & Lucas, 1971: 124, f. 26–27 (♂).
   C. s. Gerschman & Schiapelli, 1973b: 67, f. 68–73 (♂♀).
   C. s. Schmidt, 1993d: 62, f. 52–54 (♂♀).

♂♀ **stylipum** Valerio, 1982   Costa Rica, Panama
   C. s. Nentwig, 1993: 95, f. 40a–d (♂♀).

♂♀ **symmetricum** (Bücherl, 1949)   Brazil
   Magulla symmetrica Bücherl, 1950: 1, f. 1–3 (D♂).
   Magulla symmetrica Bücherl, 1957: 391, f. 36–36a (♂).
   C. s. Schmidt, 1993d: 62, f. 61–63 (♂♀).

Gen. **Cyriocosmus** Simon, 1903

♂♀ **elegans** (Simon, 1889)   Venezuela, Brazil, Bolivia
   C. e. Schiapelli & Gerschman, 1945: 181, pl. VIII (D♂).
   C. e. Schenkel, 1953a: 3, f. 3a–c (♂).
   Pseudohomoeomma fasciatum Bücherl, 1957: 391, f. 38–38a (♂).
   Pseudohomoeomma fasciatum Bücherl, Timotheo & Lucas, 1971: 125, f. 35–37
      (♂♀).
   Chaetorrhombus semifasciatus Bücherl, Timotheo & Lucas, 1971: 126, f. 42–43 (♀).

♂♀ **sellatus** (Simon, 1889)   Brazil
   C. s. Gerschman & Schiapelli, 1973b: 67, f. 16–22 (♂♀).
   C. s. Schiapelli & Gerschman, 1973b: 65, f. 1–4, 16–18 (♂♀).
   C. s. Schmidt, 1993d: 63, f. 64–67 (♂♀).

Gen. **Cyrtopholis** Simon, 1892

Transferred to other genera:
C. cyanea Rudloff, 1994 – see **Citharacanthus.**

♂   **agilis** Pocock, 1903   Hispaniola
   C. a. Bryant, 1948b: 337, f. 6–7 (♂).

Fig. 1.4. Partial listing from a systematic catalog, in this case of the spider family Therophosidae (Platnick, 1997). This catalog includes, for each taxon, information on placement in the classification, name used, sexes known, figures published, distributional information, and the literature references containing these data.

CD-ROM version of the *Index Kewensis* will become available for many groups of animals (see Chapter 8).

Theoretical Literature

Most scientific fields have technical journals that deal with the methods of the discipline. Systematics is no exception. The oldest such journal is *Systematic Biology* (originally *Systematic Zoology*) published by the Society of Systematic Biologists. This journal has included, since its inception in 1952, many important articles pertinent to both zoology and botany. *Cladistics,* published by the Willi Hennig Society, is largely a journal of papers dealing with methods, although like *Systematic Biology,* it also includes articles dealing with the classification of particular groups of organisms.

Taxon, published by the International Association for Plant Taxonomy, contains articles of general interest to systematic botanists, including "official commentary" on botanical nomenclature. *Systematic Botany,* published by the American Society of Plant Taxonomists, includes articles of broad general interest to systematists as well as articles dealing with the classification of particular plant groups.

A number of other journals such as the *Biological Journal of the Linnaean Society* and *Journal of Zoological Systematics and Evolutionary Research* (formerly *Zeitschrift für zoologische Systematik und Evolutionsforschung*), and occasional symposium or edited volumes also contain articles dealing with the theory and methods of systematics.

The burgeoning study of molecular data applicable to systematics has spawned a number of journals, including *Journal of Molecular Evolution, Molecular Biology and Evolution,* and *Molecular Phylogenetics and Evolution.* These deal mainly with the results of molecular studies and the analysis of molecular data.

Textbooks

Since the appearance of Hennig's *Phylogenetic Systematics* in 1966, a number of "essays" on systematics have appeared. Each has its own strengths, but most are somewhat dated because of advances in the field, particularly the use of numerical methods. A selection is listed at the end of the chapter under Suggested Readings.

Abstracting and Indexing Sources

If the numbers of animal taxa are at times overwhelming, so are the numbers of papers published on them. This was well recognized in the last century and was the reason for starting the *Zoological Record,* an annual digest of the literature on animals, first published in 1864. This basic reference is available in most research libraries. At the time of its inception, the organization of the *Zoological Record* contained a bibliography whose cited articles were indexed primarily for their systematic contents. Now many additional subject areas are indexed.

Biological Abstracts began publication in 1926. Unlike *Zoological Record, Biological Abstracts* publishes summaries of articles and books appearing in the primary biological literature. The major themes of published articles are also indexed.

Literature on fossil vertebrates is summarized in the works of Hay (1902), Romer (1962), and the ongoing *Bibliography of Fossil Vertebrates* published by the Society of Vertebrate Paleontology. Additional abstracting sources exist for agriculture, and other fields, and are often valuable for finding articles in the more strongly applied literature.

Computer searches of the *Zoological Record* and *Biological Abstracts* are possible beginning at about 1970. The catalogs of major research libraries, most of which now exist in database format, can also often be searched locally or via the Internet, offering additional sources of information on the systematic literature.

Systematic Collections

Systematic collections serve as repositories for specimens in the same way that libraries serve as the repositories for documents. The most important and broad-ranging natural history collections are in large public museums. Most of these are in Western Europe and North America. No museum has representatives of all species, but each has its strengths. Thus, it is part of the training and experience of systematists specializing on a given group to learn the repositories of specimens important to their work.

The *Index Herbariorum* (Holmgren et al., 1990) lists herbaria by country with a description of their collections, staff, and loan policies. Extensive indexing makes finding information on the deposition of plant collections and the specialists who work on them relatively straightforward.

The situation in zoology is much more complicated. Although published directories exist for workers in some groups, many such directories are badly out of date. Likewise, there are few comprehensive published listings of museum holdings. Thus, students in zoology wishing to learn who is working on a certain group or where relevant material is deposited should first consult their librarian or a specialist in the group for resource guides. If no useful sources are found, information must be acquired on a more piecemeal basis, through perusal of the primary literature, experience, and word of mouth.

We might expect to see dramatic changes in our ability to acquire information on collections for both botany and zoology in the coming years. Partial lists of specimen holdings are now available on the World Wide Web for some collections, and the number is increasing on a daily basis. Thus, a computer and access to the Internet may soon allow for discovery of sources on a much more rapid and efficient basis than has heretofore been possible.

Literature Cited

Bock, W. 1974. Philosophical foundations of classical evolutionary classification. *Syst. Zool.* 22:375–392.

Bremer, K., and H.-E. Wanntorp. 1978. Phylogenetic systematics in botany. *Taxon* 27:317–329.

Brundin, L. 1966. Transantarctic relationships and their significance, as evidenced by chironomid midges. *Kungl. Svenska Vetenskapsakademiens Handlingar.* Fjarde series, 11:1–472.

Cracraft, J. 1988. Speciation and its ontology: The empirical consequences of alternative concepts for understanding patterns and processes of differentiation. pp. 28–59. *In:* Otte, D., and J. A. Endler (eds.), *Speciation and its Consequences.* Sinauer Associates, Inc., Sunderland, Massachusetts.

Cracraft, J. 1992. The species of the birds-of-paradise (*Paradisaeidae*): Applying the phylogenetic species concept to a complex pattern of diversification. *Cladistics* 8:1–43.

Crowson, R. L. 1970. *Classification and Biology.* Aldine Publishing Co., Chicago. 350 pp.

Ereshefsky, M. (ed.). 1992. *The Units of Evolution: Essays on the Nature of Species.* The MIT Press, Cambridge, Massachusetts. 405 pp.

Eschmeyer, W. N. 1990. *Catalog of the Genera of Recent Fishes.* California Academy of Sciences, San Francisco. 697 pp.

Farris, J. S. 1971. The hypothesis of nonspecificity and taxonomic congruence. *Ann. Rev. Ecol. Syst.* 2:277–302.

Futuyma, D. J. 1998. *Evolutionary Biology,* Third Edition. Sinauer Associates, Sunderland, Massachusetts. 751 pp.

Hay, O. P. 1902. Bibliography and Catalog of Fossil Vertebrata of North America. *Bull. U. S. Geol. Surv.* No. 179, 868 pp.

Hennig, W. 1966. *Phylogenetic Systematics.* University of Illinois Press, Urbana. 263 pp.

Holmgren, P. K., N. H. Holmgren, and L. C. Barnett. 1990. *Index Herbariorum. Part I. The Herbaria of the World.* Eighth Edition. New York Botanical Garden. 693 pp.

Hull, D. 1970. Contemporary systematic philosophies. *Ann. Rev. Ecol. Syst.* 1:19–54.

Index Kewensis. CD-ROM Version. Oxford University Press, Oxford.

The Kew Record of Taxonomic Literature Relating to Vascular Plants. 1971– H. M. Stationery Off., London.

Lawrence, G. H. M. 1951. *Taxonomy of Vascular Plants.* Macmillan Company, New York. 823 pp.

Loconte, H., and D. W. Stevenson. 1990. Cladistics of the Spermatophyta. *Brittonia* 42:197–211.

Loconte, H., and D. W. Stevenson. 1991. Cladistics of the Magnoliidae. *Cladistics* 7:267–296.

Mabberley, D. J. 1997. *The Plant Book: A Portable Dictionary of the Higher Plants.* Cambridge University Press, Cambridge. 707 pp.

Mayr, E. 1982. *The Growth of Biological Thought: Diversity, Evolution, and Inheritance.* Belknap Press of Harvard University Press, Cambridge, Massachusetts.

Mayr, E., E. G. Linsley, and R. L. Usinger. 1953. *Methods and Principles of Systematic Zoology.* McGraw-Hill Book Company, New York.

Mitchell, P. C. 1901. On the intestinal tract of birds; with remarks on the valuation and nomenclature of zoological characters. *Trans. Linnaean Soc. London,* Zool. ser. 2, 8:173–275.

Nelson, G., and N. Platnick. 1981. *Systematics and Biogeography. Cladistics and Vicariance.* Columbia University Press, New York. 567 pp.

Platnick, N. I. 1997. *Advances in Spider Taxonomy 1992–1995.* New York Entomological Society, New York. 976 pp.

Platnick, N. I., and H. D. Cameron. 1977. Cladistic methods in textual, linguistic, and phylogenetic analysis. *Syst. Zool.* 26:380–385.

Romer, A. S. 1962. *Bibliography of Fossil Vertebrates Exclusive of North America. 1509–1927.* Mem. Geol. Soc. Amer., No. 87, 2 vols., 1544 pp.

Ross, H. H. 1974. *Biological Systematics.* Addison-Wesley Publishing Company, Inc. Reading, Massachusetts. 345 pp.

Sibley, C. G., and B. L. Monroe, Jr. 1990. *Distribution and Taxonomy of Birds of the World.* Yale University Press, New Haven, Connecticut. 1111 pp.

Simpson, G. G. 1961. *Principles of Animal Taxonomy.* Columbia University Press, New York. 247 pp.

Sokal, R. R., and P. H. A. Sneath. 1963. *Principles of Numerical Taxonomy.* W. H. Freeman and Company, San Francisco. 359 pp.

Standley, P. C. 1922. *Trees and Shrubs of Mexico.* Contribution of the United States Herbarium, Vol. 23. Smithsonian Institution, Washington, D.C.

Stonedahl, G. M. 1988. Revisions of *Dioclerus, Harpedona, Mertila, Myiocapsus, Prodromus,* and *Thaumastomiris* (Heteroptera: Miridae, Bryocorinae: Eccritotarsini). *Bull. Amer. Mus. Nat. Hist.* 187:1–99.

Stonedahl, G. M. 1990. Revision and cladistic analysis of the Holarctic genus *Atractotomus* Fieber (Heteroptera: Miridae: Phylinae). *Bull. Amer. Mus. Nat. Hist.* 198:88 pp.

Wielgorskaya, T. 1995. *Dictionary of the Generic Name of Seed Plants.* Consulting editor Armen Takhtajan. Columbia University Press, New York. 570 pp.

Willis, J. C. 1973. *A Dictionary of the Flowering Plants and Ferns.* Eighth Edition. Revised by H. K. Airy Shaw. Cambridge University Press, Cambridge. 1245 pp.

Wilson, D. E., and D. M. Reeder (eds.). 1993. *Mammal species of the World: A Taxonomic and Geographic Reference,* second edition. Smithsonian Institution Press, Washington, D.C. 1206 pp.

Zimmermann, W. 1943. Die Methoden der Phylogenetik, pp. 20–56. *In:* Heberer, G. (ed.), *Die Evolution der Organismen.* Jena.

Suggested Readings

Crowson, R. L. 1970. *Classification and Biology.* Aldine Publishing Co., Chicago. 350 pp. [Somewhat dated but still informative essay on the issues and methods of biological classification]

Eldredge, N., and J. Cracraft. 1980. *Phylogenetic Patterns and the Evolutionary Process.* Columbia University Press, New York. [Essay on cladistics and its relationship with the study of evolution]

Kitching, I. J., P. L. Forey, C. J. Humphries, and D. M. Williams. 1998. *Cladistics: The Theory and Practice of Parsimony Analysis.* Second Edition. Oxford University Press, Oxford. 228 pp.

McKenna, M. C., and S. K. Bell. 1997. *Classification of Mammals above the Species Level.* Columbia University Press, New York. 535 pp. [Part I includes comments on the history and theory of classification]

Nelson, G. 1989. Species and taxa. pp. 60–81. *In:* Otte, D. and J. A. Endler (eds.), *Speciation and Its Consequences.* Sinauer Associates Inc., Sunderland, Massachusetts [A lively discussion of species and cladistics]

Nelson, G., and N. Platnick. 1981. *Systematics and Biogeography. Cladistics and Vicariance.* Columbia University Press, New York. 567 pp. [Chapter 2 is a readable and stimulating review of the history of systematic thought]

Schoch, R. M. 1986. *Phylogeny Reconstruction in Paleontology.* Van Nostrand Reinhold Company, New York. 353 pp. [Excellent literature review of cladistics, with special reference to paleontology]

Wiley, E. O. 1981. *Phylogenetics: The Theory and Practice of Phylogenetic Systematics.* John Wiley and Sons, New York. 439 pp. [An early, and still useful, textbook on the methods of phylogenetic analysis and other aspects of systematics]

2

Nomenclature

The parlance of systematic biology is rooted in the system of binomial (or binominal) nomenclature, codified in the publications of Linnaeus. Those works, *Species Plantarum* (1753) and *Systema Naturae, 10th Edition* (1758), are the chosen starting points for modern biological nomenclature in most groups of plants and animals. Concomitant with the introduction of a consistent binomial system of names, Linnaeus organized knowledge of the living world in the form of a hierarchic classification. The fundamentals of the system of binomial nomenclature and the essential aspects of the Linnaean hierarchy will be explored in this chapter.

Nomenclature is not an end, but rather a necessary adjunct to the organization of information on biological diversity. The binomial system that was first used in a uniform way in the works of Linnaeus has persisted because of its functionality, because it is the only system that has been universally accepted, and because the entire 250-year history of biological nomenclature is based on it.

Nomenclature serves as the language by which we communicate about organisms and our knowledge of them. The names themselves do not embody information, but rather relate to concepts that can be found in the literature. Some scientific names and their associated concepts, like *Homo sapiens,* may be well known to the scientific and general public. Others, like *Capsus ater,* may be meaningful only to specialists.

Latin was the language of systematics in the time of Linnaeus. The codes of nomenclature, which will be discussed below, have traditionally placed great emphasis on the formation of scientific names as if they were Latin or latinized, usually from Greek. At the time of this writing Latin is still an integral element in both botanical and zoological nomenclature.

Species in Linnaeus' system are referred to by two distinctive words, rather than by a descriptive phrase as had been the practice of many of his predecessors. All species are placed in genera. The names of genera are Latin nouns, the names of species are Latin adjectives in agreement with the nouns, or alternatively, are Latin nouns in apposition. Generic names begin with capital letters. Specific names are in lower case. Both words are italicized. Because every species must belong to some genus, the "name" of the species is therefore a binomial, as, for example:

Homo sapiens
Acacia drummondii

The individual words forming a scientific name are often referred to as the *generic epithet* and the *specific epithet.* Scientific names do not include diacritics, although they may be hyphenated.

All generic names (epithets) and specific names (epithets) have authors, the name(s) of the person(s) who first published them. This aspect of nomenclature probably causes more day-to-day confusion than any other. Inclusion of authors' names allows for more accurate tracing of the history of application of names. Scientific names with similar spellings are usually distinguishable from one another as other than misspellings when the author's name is included. For example:

Rhinacloa pallipes Reuter
Rhinacloa pallidipes Maldonado

Dates of authorship may also be included to help distinguish among names as well as to assist in locating relevant literature:

Macrocoleus femoralis Reuter, 1879
Cyrtocapsus femoralis Reuter, 1892
Psallopsis femoralis Reuter, 1901

If the name of the author is in parentheses, this indicates that a species is placed in a genus other than the one in which it was originally described. In the botanical literature, the name of the author of the species may be placed in parentheses and also be followed by a second name, the latter representing the author who moved the species to its genus of current placement:

Werneckia minuta (Werneck)
Ceratozamia boliviana Brongn.
Zamia boliviana (Brongn.) A. DC.

In zoological catalogs and other listings, names as used by persons other than the original author are often written with the name of that person separated from the scientific name by a colon, as, for example:

Phytocoris marmoratus Blanchard
Phytocoris marmoratus: Stonedahl

the latter listing indicating use of the concept by an author other than the original.

Codes of Nomenclature

Most early authors subsequent to Linnaeus followed the example of his now famous works and adopted a strictly binomial system of naming. Some did not, and their works are now largely rejected and forgotten with respect to the names they applied to plants and animals.

The discussions of nomenclature in this chapter deal with issues treated in the botanical and zoological codes. There are also codes for cultivated plants (1980) and bacteria (1976), and students of those areas should consult the appropriate volumes.

The first formal attempt to add order to the creation of names in zoology came in 1840 with the British Association Code or Stricklandian Code, named after its author, Hugh Edward Strickland. The first truly international efforts in zoology involved the establishment of the International Commission of Zoological Nomenclature, which published the *Regles Internationales de la Nomenclature Zoologique* in 1905. Since that time the rules of nomenclature have been administered and periodically modified by the International Commission. The most recent *International Code of Zoological Nomenclature* ("Zoological Code"), published in 1985, is divided into a series of articles and recommendations. A revision of the Code is in preparation at the time of this writing. "Articles" are intended to be followed strictly by those involved in the creation or modification of names, whereas "recommendations" are intended as guidelines that should be followed. The Code also contains a glossary of terms used in zoological nomenclature, a useful aid in interpreting the Code and other writings pertaining to nomenclature.

The regulation of names in botany has a history similar to that in zoology, with the adoption of a set of "laws" at the Paris International Botanical Congress in 1867. The *International Code of Botanical Nomenclature* was published in 1952, supplanting all prior codes and implementing a more orderly approach to the formation and regulation of names. Like the Zoological Code, the Botanical Code is divided into articles and recommendations. The most recent Code of Botanical Nomenclature was published in 1994. Revised editions of the Code appear at six-year intervals corresponding to the International Botanical Congresses. All recommendations for changes in the Botanical Code are published in *Taxon*.

According to its preamble, the Code of Zoological Nomenclature was designed with the intent of promoting "stability and universality in the scientific names of animals." The Botanical Code ". . . aims at the provision of a stable method of naming taxonomic groups." As we shall see elsewhere, stability for its own sake is not a desirable property, however. The names covered by the codes can be divided into the following groups. A name proposed in a given grouping can serve at any level within that grouping.

- *Suprafamilial names.* Names above the family-group level are unregulated in zoology, need not be based on generic names, and do not usually have standardized endings. In botany, it is recommended that such names be based on a nomenclatorial type (generic name) and have standardized endings.

- *Family-group names.* This group includes all names above the level of genus-group, up to and including superfamily; the most commonly included ranks are tribe, subfamily, family, and superfamily (except for the last in botany). Names in this group have standardized endings in both botany and zoology (with eight allowed exceptions in botany for names used by Linnaeus, as, e.g., Compositae and Leguminosae) and are based on generic names (see below under the Linnaean hierarchy).
- *Genus-group names.* This group includes generic and subgeneric names, to which the same rules apply.
- *Species-group names.* This group includes species and subspecies names in zoology and additional infrasubspecific names in botany, with the same rules applying at all levels.

Although the codes for zoology and botany differ in their organization, the basic tenets, stated in the form that they are found in the Zoological Code, can be outlined as follows:

1. Priority
2. Availability
3. Typification
   a. Species-group types
   b. Genus-group and family-group types
4. Homonymy
5. Synonymy

Priority

The principle of priority dictates that the first name applied to a taxon is the name that will be used. One might ask, "What is the problem? Doesn't every taxon have just one name"? The answer is "Not necessarily." Imagine, for example, different authors working in different places, not aware of each other's activities. Under such circumstances the same species may have been described from two to several times. Or, possibly more common, a given species shows great variability, which was not understood at the time of its initial discovery. The same or different authors might apply different names to the taxon, unaware that they were dealing with morphologically distinct males and females of the same species, for example.

The concept of priority sometimes runs afoul of nomenclatorial (or nomenclatural) stability when the oldest name is not the one that has been most widely used. The codes address such circumstances. The modern codes all have provisions for setting aside long unused names, even though those names might be favored by the rule of priority. In zoology, such names may be placed on a list of "rejected" names by appeal to the International Commission of Zoological

Nomenclature. Names are "conserved" in botany in deference to their long-standing usage.

As an example of applying the rule of priority:

Lygaeus saltitans Fallen was published in 1807. *Chlamydatus* Curtis was described in 1833, with a new species *marginatus* Curtis. Fieber in 1858 described a new genus *Agalliastes,* including *saltitans* (Fallen), among other species. The type of *Agalliastes* was fixed as *saltitans* by Kirkaldy in 1906. Flor synonymized *marginatus* Curtis with *saltitans* Fallen. Thus, *Chlamydatus saltitans* (Fallen) is the name that must be used on the basis of priority.

Availability

Whereas priority is a comparatively objective criterion, availability is more nebulous. In the context of the codes, most names would be considered "available" if they meet the following four criteria:

a. appear in a work published after 1753 for plants and 1758 for most animals;
b. meet the criteria for 'publication' designated in the codes;
c. are written in the Latin alphabet (in this day and age the English alphabet); and
d. are binomial (if referring to species).

The codes also require or recommend that newly proposed names be accompanied by a description of the taxon (the text of which must be written in Latin in botany) and have a designated type (see below).

Of these four criteria, publication is the most difficult to circumscribe.

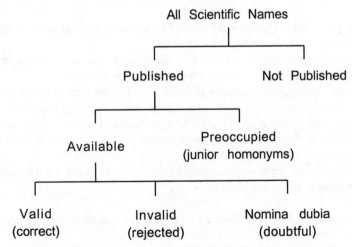

Fig. 2.1. The categories of scientific names (modified from Blackwelder, 1967).

Although many works from the past have presented problems with regard to satisfying the criteria of publication, no traditional printed works present the difficulties being encountered in the age of photocopying and computers. For example, does a dissertation issued through University Microfilms constitute a publication? Are works that appear only on the World Wide Web published in the sense of the codes? Such questions are currently being addressed, with the aim of making the nomenclatorial transition into the electronic age an orderly one.

Typification

Whereas the concepts of priority and availability were more-or-less implicit in the works of many authors from the time of Linnaeus, such was not the case with the type concept. To early authors, the most common species was often considered "typical" of a genus. Under this approach a landmark for fixing generic limits was often recognized, but the absence of any formalized approach to such fixation at times created great confusion. Species were usually represented by a number of "typical" specimens.

The type concept was not formally codified in zoological nomenclature until the publication of the *Regles* (the first Zoological Code) in 1905. The required designation of "types" brought increased stability and order to zoological nomenclature, particularly in the application of generic names. The mandatory designation of types in botany did not apply until January 1958. Types are of two forms, names and specimens.

Species-group types. A type in the species group is a specimen to which a name is attached, providing an objective criterion for establishment of usage of that name. Species-group types recognized in the codes of nomenclature are called *primary types* and are of the following sorts:

- *holotype:* A single specimen designated by the author of the name at the time of publication of the original description.
- *neotype:* A specimen designated to replace the holotype (or other primary type) if the latter can be documented as lost or destroyed.
- *syntypes:* A group of specimens thought to represent a given species, as designated or indicated by the author(s) of the original description. These specimens are sometimes referred to as the 'type series.' The term *co-type* is sometimes used in the same sense. If such specimens exist, only from them may a lectotype be designated.
- *lectotype:* One of the syntypes chosen by the original or a subsequent author(s) to function as the name bearer.

It is customary to deposit primary types in a recognized institution dedicated to the maintenance of scientific collections.

Many other categories of so-called types are mentioned in the literature. In botany, probably most important among these is the *isotype,* a specimen collected from the same individual plant as the holotype. Such an approach in zoology is possible in the case of colonial organisms, but the term isotype is not applied.

Paratypes are commonly designated in conjunction with the description of new species, these being specimens that were studied by the author of a new species or subspecies and designated by that person at the time of publication of the original description. Although paratypes often have value to subsequent workers for reference as to how a name was applied, they have no standing in nomenclature and, as one might imagine, sometimes include misidentified specimens, as is also often the case with syntypes. Paratypes may serve as convenient potential candidates for designation as neotypes, in cases where holotypes are destroyed.

Genus-group types. These types are species, comprising names, not specimens. Classical authors generally did not designate types of genera. The modern codes require that for a generic or subgeneric name to become available, a type species must be designated by the author. Types for genera published before institution of requirements for mandatory designation in the codes are most commonly *fixed,* that is, assigned, by either monotypy or subsequent designation:

- *monotypy:* The genus in question was published with only a single included species, in which case the genus would have been *monotypic* (or *monobasic*) when made available; only the single originally included species can be the type even though other species may have been added to the genus subsequently; or,
- *subsequent designation:* When more than one species was originally included, the type is selected from them at a later time either by the original author or another person.

Family-group types. Types in this group are genera. Family-group names are subject to the rules of homonymy (see below), in that no two can be spelled identically, even though they may be based on different, although similar, generic names. Just because the generic name on which the oldest family-group name is based is in synonymy (see below), this does not invalidate the family-group name. Examples may help to illustrate the point. In the insect suborder Heteroptera, the family Velocipedidae contains a single valid genus *Scotomedes.* The generic name on which the family name is based, *Velocipedes,* is a junior synonym of *Scotomedes.* The angiosperm genus *Winteria* is a junior synonym of *Drimys,* yet the valid family name is Winteraceae.

Homonymy

The principle of homonymy refers to the application of the same name to different taxa. The codes of nomenclature state that no two names above the species-group level may be the same in either zoology or botany, although names may be duplicated between the two fields. This dictate has been applied since early on, with the greatest confusion having resulted from names with variant spellings but which nonetheless represent the same word or meaning. Currently in zoology, the rules allow differently spelled names to be treated as different even though they may be based on the same root and have the same intended meaning. Homonyms may be of several types.

Senior homonym. The available name on the basis of priority.

Junior homonym. The preoccupied name (not in use) on the basis of priority or by ruling of the bodies governing nomenclature.

Primary homonym. In the species group (species, subspecies, etc.), these are names that are the same and were proposed in the same genus-group taxon. The junior homonym must always be replaced, either by a new name or by a junior synonym, if such exists.

Secondary homonym. If two species with the same name are placed in the same genus subsequent to their publication, they become secondary homonyms. The senior secondary homonym is the older of the two (or more) names. The most obvious difficulties in adjudicating questions of secondary homonymy will be encountered when the rules of Latin are applied strictly. For example, species that are black are often named for their coloration. The spelling of the name will vary depending on the gender of the genus: *niger* (masculine), *nigra* (feminine), or *nigrum* (neuter). If two species named for their black coloration end up in the same genus, a homonym will be formed under the strict application of the rules of Latin grammar, even though the spellings were different because the genera of original placement were of different gender. Additional complications arise from orthographically incorrect constructions such as *nigrus,* whose intended meaning was "black." The replacement of secondary homonyms may vary depending on dates of publication and between botany and zoology. Therefore the codes should be consulted on these issues.

The following examples will help illustrate the rules of homonymy:

- *Bougainvillia* Lesson (Coelenterata), is not a homonym of *Bougainvillea* Comm. ex Juss. (Angiospermae: Nyctaginaceae)
- *Kingia* Malloch, 1921 (Insecta: Culicidae) is not a homonym of *Kingia* Brown (Angiospermae: Xanthorrhoeaceae), but is a junior homonym of *Kingia* Theobald, 1910 (Insecta: Culicidae).
- *Mononychus* Perle, Norell, Chiappe, and Clark, 1993 (Vertebrata: Dinosauria), is a junior homonym of *Mononychus* Scheuppel, 1824 (Insecta: Coleoptera)

- *Gerris* Fabricius, 1794 (Insecta: Heteroptera) and *Gerres* Quoy and Gaimard, 1824 (Pices), are not homonyms, in spite of the similar spellings. The family names based on these genera are now spelled differently, Gerridae and Gerreidae, respectively, to avoid homonymy.
- *Phytocoris marmoratus* Douglas and Scott, 1869 (Insecta: Heteroptera), is a junior primary homonym of *Phytocoris marmoratus* Blanchard, 1852, both having been placed in the same genus at the time of their original description.
- *Phytocoris modestus* Reuter, 1908, is a junior primary homonym of *Phytocoris modestus* Blanchard, 1852, even though *modestus* Blanchard is currently placed in the genus *Polymerus*.
- *Dichrooscytus marmoratus* Van Duzee, 1910, became a junior secondary homonym of *Phytocoris marmoratus* Blanchard, 1852, on the basis of priority, when it was transferred to *Phytocoris*.

Synonymy

The concept of synonymy in nomenclature deals with the application of different names to the same taxon. As is the case with homonyms, synonyms can be of several types.

Senior synonyms. The senior name is one of two or more different names that is deemed valid, usually on the basis of priority or because of its choice by the first reviser (in zoology), or by a ruling of the bodies governing nomenclature.

Junior synonyms. The junior name(s) is the one deemed to be invalid, usually on the basis of priority, or because of its choice by the first reviser, or by a ruling of the bodies governing nomenclature.

Objective synonyms. These are different names that by examination of nomenclature alone can be judged to refer to the same taxon. For example, any two family-group names with the same type genus or any two genera with the same type species are objective synonyms; two species based on the same specimen would also be objective synonyms. Because of their clear-cut nature, objective synonyms are usually created only by inadvertent error. For example:

> *Saprophilus* Steubel, 1939 (Insecta: Coleoptera), is a junior objective synonym of *Creophilus* Leach, 1819, both being based on the same type species, *Staphylinus maxillosus* Linnaeus, 1758.

Subjective synonyms. These are different names whose application to the same taxon is determined by a systematist. For example, two nominal species originally described as distinct but later treated as being identical are subjective synonyms. This type of synonymy is common, and in some complex cases the subject of great confusion.

Demarata Distant, 1884 (Heteroptera: Miridae), is a junior subjective synonym of *Ceratocapsus* Reuter, 1876, because the type species of *Demarata, D. villosa* Distant was judged, subsequent to its description, to belong to *Ceratocapsus*.

The Linnaean Hierarchy

The Linnaean hierarchy (Fig. 2.2) is a system in which the names associated with the levels connote rank. The same result could be achieved through a system of indentation, the degree of indentation connoting rank in the hierarchy. Indentation requires a visual presentation in order to understand rank. The Linnaean system of names does not — one need only remember the rank order of the names used to define the hierarchy. Linnaeus had a limited number of levels (ranks) in his system, but he nonetheless reflected the nesting of groups within groups. The systematic hierarchy in the 10th edition of the *Systema Naturae* (1758) included:

Kingdom
  Class
    Order
      Genus
        Species
          Variety

Later authors incorporated more categories above and below the level of order. McKenna and Bell (1997) discussed this issue at length with respect to the Mammalia and probably incorporated as many levels as does any existing, formal, modern classification. Today it is common to see ten or more levels in both botany and zoology.

Kingdom
  Phylum
    Class
      Subclass
        Order
          Suborder
            Superfamily (primarily in zoology)
              Family
                Subfamily
                  Tribe
                    Genus
                      Species

INSECTA COLEOPTERA. Scarabæus. 345

I. COLEOPTERA.

Elytra alas tegentia.

170. SCARABÆUS. *Antennæ* clavatæ capitulo
fiſſili.
Tibiæ anticæ ſæpius dentatæ.

* *Thorace cornuto.*

Hercules. 1. S. thoracis cornu incurvo maximo ſubtus barbato, ca
pitis cornu recurvato: ſupra dentato.
Marcgr. braſ. 247. *f.* 3. *Jonſt. inſ. t.* 16. *f.* 1.
Olear. muſ. t. 16. *f.* 1. *Pet. gaz. t.* 70. *f.* 1.
Grew. muſ. 162. *Swamm. bibl. t.* 30. *f.* 2.
Rœſ. ſcarab. 1. *t. A. f.* 1. *inſ.* 4. *p.* 45. *t.* 5. *f.* 3.
Habitat in America.

Actæon. 2. S. thorace bicorni, capitis cornu tridentato: apice bi-
fido. *Muſ. L. U.*
Marcgr. braſ. 246. Enena. *Olear. muſ. t.* 16. *f.* 2.
Mer. ſur. t. 72. *Rœſ. ſcar.* 1. *t. A. f.* 2.
Hoffn. pict. 1. *t.* 1. *in medio. Swamm. bibl. t.* 30. *f.* 4.
Habitat in America.

Şimſon. 3. S. thorace bicorni, capitis cornu apice tantum bifurcato.
Sloan. jam. 2. *p.* 205. *t.* 237. *f.* 4. 5.
Brown. jam. 428. *t.* 43. *f.* 6. Scarabæus 4.
Habitat in America.

Atlas. 4. S. thorace tricorni: antico breviſſimo, capitis cornu re-
curvato. *M. L. U.*
Marcgr. braſ. 247. *f.* 1. *Olear. muſ. t.* 16. *f.* 3.
Pet. gaz. t. 49. *f.* 8. *an t.* 14. *f.* 12.
Merian. ſurin. in titulo F. G. Swamm. bibl. t. 30. *f.* 3.
Habitat in America.

Aloëus. 5. S. thorace tricorni: intermedio longiore, capite muti-
co, elytris uniſtriatis. *M. L. U.*
Rœſ. inſ. 2. *ſcar.* 1. *t. A. f.* 6. *Pet. gaz. t.* 24. *f.* 10.

Y 5 *Habi-*

Scarabæorum *Larvæ vivunt tranquillæ ſub terra; harum pleræque fimo delectantur
& eo paſcuntur.*

Fig. 2.2. The Linnaean hierarchy as portrayed by a page from the *Systema Naturae,* including the ranks of Class (Insecta), Order (Coleoptera), Genus (*Scarabaeus*), and species (*Hercules,* etc.) (Linnaeus, 1758).

Criticisms of the Linnaean System

Criticisms of the Linnaean hierarchy as embodied in the codes of nomenclature can be divided into two general categories. Both address the issue of stability, a caveat of the codes.

First, is modification of names themselves, irrespective of the taxonomic concepts attached to them. The most obvious example is the creation of a *new combination.* When a species is transferred from one genus to another, and the gender of the generic names is different, the termination of the specific name may have to be changed to maintain gender agreement. For example:

Dichrooscytus speciosus Van Duzee (Insecta: Heteroptera) became *Bolteria speciosa* (Van Duzee) in order to maintain gender agreement upon its transfer to *Bolteria.*

Gender agreement has been a particular irritant to zoologists, where the code dictates the gender of many generic names, irrespective of the gender implied in their usage by the original author. The result has been that some "authorities" on nomenclature have modified specific names just to achieve agreement between the gender of the genus — as determined by the code — and all of its included species.

The requirement of gender agreement is an impediment to stability of species-group names that could be removed by not requiring agreement between the nouns and their adjectives. Such an approach would require deciding whether species should adopt their original termination or their current termination. It would also remove strict interpretation of the rules of Latin grammar in the creation of names in the future. The consistent spelling of names has particular appeal for the use of computer databases, which are powerful tools for the organization and management of information in biological systematics.

One might therefore expect adherence to the rules of Latin grammar to come under more pressure for change. But, in the most recent, yet unpublished, revision of the Zoological Code, gender agreement is maintained, in spite of efforts to remove it. It should be pointed out that incorrect Latin orthography does not affect the availability of names in zoology. Such bastard names have usually been treated as if they were simply an arbitrary combination of letters, a legitimate approach under the Zoological Code. As an example we might point once again to the erroneous use of *nigrus* as the masculine form for reference to species that are black in color.

Second, and less subject to objective resolution, is the issue of the systematic concepts attached to names. Typification brought a sense of order to nomenclature that had not been achieved previously, this "transition" occurring earlier in zoology than in botany. Yet, it poses what some systematists have viewed as problems with no obvious or universally acceptable solution, especially in light of the long nomenclatorial history with which biologists must contend.

At the simplest level, consider type specimens. A poorly chosen specimen for a given species may leave the identity of the taxon in permanent doubt. For example, a female holotype in a genus in which species identification is based on details of male genitalia may render comparisons between this and other species moot if positively associated males of the female-based species cannot be found. Nonetheless, in the vast majority of cases, the existence of type specimens — in combination with well-prepared printed descriptions and figures — has greatly facilitated the uniform application of species names.

All supraspecific categories have "nomenclatorial" types. In the case of genera, a "poorly chosen" type species can cause problems with generic concepts. In the example above, if the designated female holotype pertains to the type species of the genus, the issue of generic relationships may be complicated or unresolvable.

Another implication of typification is sometimes manifested in groups with histories that predate the establishment of the codes. When, upon revision or other critical examination, two long-used and well-known generic names are found to be synonymous because their type species are deemed to be congeneric, one of the generic names will be treated as invalid under the rule of priority. Such a situation would be of no import to a computer, but it has proved vexing for taxonomists and other biologists who attach particular concepts to certain names. The last line of recourse is to petition the international commissions for conservation of the name whose concept is more widely understood, thus setting aside the rule of priority.

If fixed types were removed from the equation, concepts could probably be made more stable by changing the type of a supraspecific taxon — from time to time — to improve conceptual agreement. This would, however, leave the field wide open for changes based on the personal choice of investigators with competing theories of generic limits and relationships. The choices would range from placing all species in a single genus, on one extreme, to describing a unique genus for each species, on the other.

Thus, the issue seems to boil down to whether maintaining priority and typification are the best possible solutions in a situation that, under any circumstances, will always involve some compromise. At least two alternative approaches have been proposed.

Uninomial nomenclature (Michener, 1963) was designed to do nothing more than promote stability of names. It received little attention at the time of its introduction into the literature and now seems to have disappeared from consideration.

More recent is the proposal of *phylogenetic taxonomy,* which — unlike uninomial nomenclature — derives directly from philosophical constructs, adopted almost without modification from a disquisition by Hull (1965) on the influence of Aristotelian essentialism on taxonomy. We will discuss the substance of Hull's work in Chapter 3. Because the implications of phylogenetic taxonomy are more systematic than nomenclatorial, that subject will be taken up in Chapter 8.

Aids to the Use of Nomenclature

With at least 2.5 million described species of living organisms, the problems of finding and applying names correctly requires some well-organized aids. These vary between botany and zoology, and will therefore be discussed separately.

Zoology

The International Commission on Zoological Nomenclature issues two publications. First is the Code, discussed above, which has undergone several revisions, the most recent available published in 1985, but with an update scheduled for release late in 1999. For any working taxonomist, a copy of the most recent code, and doubtless older versions, is indispensable. Second is the *Bulletin of Zoological Nomenclature,* in which petitions to the Commission and decisions of the Commission appear. Petitions include proposals for setting aside the rule of priority in order to suppress long unused names in favor of more recently published names that have been widely used and accepted by the scientific community. The Commission might also be asked to decide which of two homonymous family-group names should remain unchanged and which should be replaced.

The *Zoological Record* prepares, on a yearly basis, a listing of many of the generic (and subgeneric) names newly published in zoology. The *Nomenclator Zoologicus* (Neave, 1939–1996) includes an alphabetical listing of most generic names published in zoology through 1994.

Listings of species names are difficult to prepare on a comprehensive basis because of the multitude of taxa. Nonetheless, at least one work attempted such a universal listing for animal names published between 1758 and 1850 (Sherborne, 1902–1932). This work is valuable in discovering and attributing older names, especially those placed in large, poorly defined genera of early authors and which may now reside in several families.

Botany

Botanical systematics and its attendant nomenclature has a more streamlined history than zoology. Sources such as the *Index Kewensis* offer a centralized source of information of a sort found widely scattered in the zoological literature. On the other hand, the *Zoological Record* serves as a more-or-less comprehensive indexing source in zoology, whereas no equivalent publication exists in botany.

Literature Cited

Blackwelder, R. E. 1967. *Taxonomy: A Text and Reference Book.* John Wiley and Sons, New York. 698 pp.

Hull, D. L. 1965. The effect of Essentialism on taxonomy: Two thousand years of stasis. *Brit. J. Phil. Sci.* 15:314–326; 16:1–18 (reprinted in: Ereshefsky, M. [ed.]. 1992. *The Units of Evolution. Essays on the Nature of Species.* MIT Press, Cambridge, Massachusetts. 405 pp.)

International Code of Botanical Nomenclature. 1994. Koeltz Scientific Publishers, Koenigstein, Germany.

International Code of Nomenclature of Bacteria. 1976. American Society of Microbiology, Washington, D. C. 180 pp.

International Code of Nomenclature for Cultivated Plants. 1980. American Horticultural Society. 32 pp.

International Code of Zoological Nomenclature. 1985. International Trust for Zoological Nomenclature. The Natural History Museum, London, and University of California Press, Berkeley. 338 pp.

Linnaeus, C. 1753. *Species Plantarum.* Stockholm.

Linnaeus, C. 1758. *Systema Naturae.* 10th Edition. Stockholm.

McKenna, M. C., and S. K. Bell. 1997. *Classification of Mammals above the Species Level.* Columbia University Press, New York. 535 pp.

Michener, C. 1963. Some future developments in taxonomy. *Syst. Zool.* 12:151–172.

Neave, S. A. 1939–1996. *Nomenclator Zoologicus.* 9 vols. Zoological Society of London, London.

Sherborne, C. D. 1902–1932. *Index Animalium.* 9 volumes. British Museum (Natural History), London.

Suggested Readings

Blackwelder, R. E. 1967. *Taxonomy: A Text and Reference Book.* John Wiley and Sons, New York. 698 pp.

Borror, D. J. 1960. *Dictionary of word roots and combining forms.* Mayfield Publishing Company, Palo Alto, California. 134 pp.

Brown, R. W. 1956. *Composition of Scientific Words.* Published by the author. [A useful compendium of Latin used in biological nomenclature]

Crowson, R. A. 1970. *Classification and Biology.* Aldine Publishing Company, Chicago. [A pointed disquisition on the codes of nomenclature and their history, as well as a readable and cogent essay on biological classification]

International Code of Zoological Nomenclature. 1999. Fourth Edition. International Trust for Zoological Nomenclature. London. 306 pp.

Stearn, W. T. 1992. *Botanical Latin.* History, Grammar. Syntax, Terminology, and Vocabulary. Fourth edition, revised. Timber Press, Portland, Oregon. 546 pp.

3

Systematics and the Philosophy of Science

Objective Knowledge versus Truth in Science

Science, including systematics, deals with the acquisition of knowledge concerning the natural world. Yet, the methods of science are not directly concerned with assessing how we judge the truthfulness of the knowledge we acquire, what ideas represent mere conjecture and which represent knowledge, and whether or not we can we trust our observations of the real world. Philosophy as a field of study may help us to understand these issues. Some would say that within philosophy it is epistemology, the inquiry into the origin, nature, and limits of knowledge, that defines science. Thus, it is to this area that we now turn.

The terms *fact* and *theory* are often juxtaposed in common parlance, the former connoting a kind of certainty that the latter does not have. Ideas in science are often referred to as *theories,* or *hypotheses,* terms whose meanings will be used interchangeably in this book, even though *theory* is often used in reference to more general concepts than is *hypothesis.* Theories and hypotheses are clearly ideas to be judged. Nonetheless, the term "truth" is often used in the popular press when referring to scientific statements, and it is sometimes encountered in the writings of scientists — and philosophers. Science produces what might be referred to as "objective knowledge." The question of whether such knowledge is, or ever can be, absolutely truthful is beyond the scope of the present work. We might observe, however, that belief in absolute truth would seem to involve an act of faith, an approach to the acquisition of knowledge that science does not admit. Thus, within the present work, all scientific statements will be treated as hypothetical, albeit some of them receiving stronger support from the available evidence than others.

Some examples may help put this issue of absolute truth into perspective. Few people in this day and age would doubt the correctness of the heliocentric, or Copernican, theory of our solar system. Indeed, most would consider the idea that the earth and other planets in our solar system revolve around the sun to be irrefutable. Nonetheless, the overly complex Ptolemaic theory in which the earth was fixed in the center of the universe with the stars revolving around it was widely believed until the sixteenth century, even after many centuries of careful observation.

Within biology, it is commonly said that biological evolution has been shown to be a fact. Few — except those believing in divine creation — would deny that life on earth has evolved. Yet, the exact mechanisms allowing for the diversity of forms we see in the fossil record and living today are still the subject of considerable controversy. The "fact" of evolution may appear manifest in the changing representation of taxa found in the fossil record through time and seems to be further confirmed by the hierarchic relationships observable among taxa, both living and fossil. But, even though the generalities of these issues may appear to be "factual," our detailed understanding of them — both phylogenetic and mechanistic — is far from incontrovertibly resolved.

Consider for the moment the issue of bird–dinosaur relationships. It is often said that dinosaurs went extinct at the end of the Cretaceous period, some 65 million years ago. Yet, a widely accepted current theory based on cladistic studies indicates that birds find their closest relatives in the Theropoda, a subgroup of saurischian dinosaurs, suggesting extinction only of dinosaur lineages other than birds (Dingus and Rowe, 1998). Thus, what was not so long ago thought to be a fact, is now widely regarded as a rejected theory. Dinosaurs are not extinct — they exist as birds!

Knowledge of issues such as whether life has evolved — and how — and whether all dinosaurs went extinct 65 million years ago is not acquired simply as the result of observation. Evolution is not something we are observing on a daily basis, nor were we born with a preformed concept of dinosaur. These ideas represent scientific theories that devolved from more than two centuries of concerted observation, analysis, and synthesis on the part of biologists, geologists, and others interested in the natural world.

Prior to the early 1970s, most of the literature on *systematics* made no mention of the terms hypothesis or theory and said little about what a systematic theory might be. Neither did that literature suggest how theories might be tested or evaluated. Most systematists have probably never concerned themselves directly with issues related to the philosophy of science. Gaffney (1979a) prepared one of the first useful expositions on the subject, one that is still timely and eminently readable.

It has been said that systematics is something you do, not something you think about. This anti-intellectual viewpoint does not justify ignoring the potential importance of inquiry into the scientific thought process itself, however. The philosophy of science, nonetheless, might have retained its apparent irrelevance to the field of systematics had it not been for the application — in the middle of the twentieth century — of two divergent philosophies, or methodologies, for the practice of systematics, as outlined in Chapter 1.

With regard to the role of methods in science, Platnick (1979, 538) observed:

It may seem paradoxical that systematics (or any science) must adopt methods without being able to attest to their efficacy. But the fact is that we use our methods in an attempt to solve problems. If we already knew the correct solutions to those problems, we could easily evaluate and choose among various methodologies: those methods which provide the correct solutions would obviously be preferred. But of course, if we already knew the correct solutions, we would have no need for the methods.

The point of view espoused by Platnick seems not to be universally appreciated nor accepted by those interested in the study of systematics. For this reason we will examine some of its general and specific implications.

Scientific Theories and the Concept of Falsifiability

Approaches to science are often divided into the inductive and the deductive. The *inductive* approach holds that scientific knowledge accumulates from repeated observation of the facts. Induction might be defined, alternatively, as a process in which scientific inference leads from the specific to the general. This approach, usually attributed to the sixteenth-century English philosopher Francis Bacon, has long been considered a basic method of science. The noted philosopher of science Karl Popper (1968) quoted Reichenbach, also a philosopher, as holding the view that "the principle of induction is unreservedly accepted by the whole of science and that no man can seriously doubt this principle in everyday life either." Similar expressions of belief are widely scattered in the philosophical, scientific, and popular literature.

The perspective that "elementary statements of experience" and "judgments of perception" are to be treated as scientific is labeled by Popper as *positivist* and is equivalent to *induction.* In this inductivist or positivist perspective, observations are treated as facts. By extrapolation, if a large number of elementary observations are judged to be true, then statements that summarize general agreement among them must also be true. We have referred to positivism as exemplified by the phenetic school of thought as 'extreme empiricism' in Chapter 1.

The *deductive* approach, in contradistinction to the inductive approach, holds that all observation should be construed as theoretical or, stated another way, that all observations are theory laden. To paraphrase Rieppel (1988), observation and the search for lawful regularities can never be unbiased. Observation must be based on some expectation; research is expected to reveal something of significance, something worth recording. That significance can only derive from some theoretical context.

The hypothetico-deductive view of science maintains that the growth of knowledge derives from the formulation of theories that imply certain predictions (deductions) that can be judged by observation. These observations represent tests of the theory. The observations that conform to the deductions are said to *corroborate* the theory; those that do not are said to *falsify* it. The greater and more severe the number of tests passed, the greater the degree of corroboration.

A major distinction between the inductive and deductive approaches would seem to revolve around the concept of "truth" and how we can determine what is true and what is not. In his attempts to solve what he called the "problem of induction" Popper reasoned that 'falsifiability' was the only suitable criterion of demarcation between science and nonscience because logically there was no way to decide whether a given statement was absolutely truthful. No number of singular statements will prove the truth of a universal statement. For example, we could not be certain that the statement "all insects have compound eyes" was absolutely true, no matter how many observations we might have made, because we can never observe all insects in all of time. However, the absolute truthfulness of the statement could easily be rejected by finding a single insect without compound eyes. Thus, even if we could all agree on the truthfulness of our observations — the singular statements — we would still not be justified in believing that a more general or "universal" statement is true. Instead, one nonconforming observation or singular statement could suggest that the universal statement was false.

The issue of whether systematic statements are strictly universal or not, and whether they can therefore be validly tested using the methods described by Popper — as some philosophers would demand — is an issue that has been much discussed, but certainly not resolved in the form of anything that could be called a consensus. Many biologists believe that the requirement of universality is largely irrelevant, and certainly Popper, who was well aware of the application of his philosophy of science in systematic biology, never claimed that it was required.

The deductive method of testing, which has also been called the "falsificationist approach to science," suggests that theories can be tested in the following ways (Popper, 1968):

1. Logical consistency, whereby the conclusions are compared among themselves;
2. Investigation of the logical form of the theory, to determine whether in fact it is an empirical or scientific theory;
3. Comparison with other theories to determine whether a given theory would represent a scientific advance; and
4. Empirical testing of conclusions (deductions) from the theory.

Empirical testing is probably the most important for our purposes, although the others are also relevant. Failure of a theory to conform to empirical observation would suggest that a theory has been *falsified*. Conformity to observation would *corroborate* the theory.

Popper (1968) pointed out there is no truth in science and that one should not equate degree of corroboration with truthfulness. For example, one would not say that "a theory is hardly true so far, or that it is still false," simply because it had not been rigorously tested. Popper also rejected the idea that the truthfulness of a statement can be interpreted in the sense of the probability of that statement being true. According to Popper (1968) "The advance of science is not due to the fact that more and more perceptual experiences accumulate in the course of time. Nor is it due to the fact that we are making ever better use of our senses." Science is rather a matter of conjecture and refutation.

Degree of corroboration, then, in Popper's view is directly related to the degree to which a hypothesis has survived attempts to falsify it. Any hypothesis which by virtue of its formulation would conform to any and all observations, is not highly corroborated, but rather immunized at the outset from critical testing. *Degree of corroboration is a relative quantity, the choice being among hypotheses, not the degree to which a given hypothesis is corroborated in its own right.*

In applying the deductive approach, we might adopt the following methodological rules as outlined by Popper in *The Logic of Scientific Discovery:*

1. Science is in principle without end. Once you decide on the absolute truth, you retire from the game.

   Of course, the individual scientist may wish to establish his [or her] theory rather than refute it. But from the point of view of progress in science, this wish can easily mislead him [or her]. Moreover, if he does not himself examine his favourite theory critically, others will do so for him. The only results which will be regarded by them as supporting the theory will be the failures of interesting attempts to refute it; failures to find counter-examples where such counter-examples would be most expected, in light of the best of the competing theories (Popper, 1975, 78).

2. A corroborated hypothesis should not be dropped without reason, that is, without a stronger competing hypothesis.
3. Parsimony should be applied in the testing of hypotheses, that is, *ad hoc assumptions* [special pleading] designed to dispose of observations that would otherwise provide evidence against a theory should be avoided.

It is the last of the rules that has assumed greatest importance in modern systematic practice. We will therefore discuss the idea in it own right.

Parsimony and Ad Hoc Hypotheses

Parsimony as an approach to evaluating observations is often attributed to William of Occam (or Ockham), English philosopher of the early 1300s, and referred to as Occam's Razor. It is a widely appreciated approach in science wherein the number of assumptions required to explain observations is minimized. Assumptions invoked over and above the minimum number required are often referred to as "ad hoc." The use of parsimony was obvious in the original formulation of phylogenetic methods by Hennig, although Hennig never used the term and did not refer to Occam's Razor. As pointed out by Farris, parsimony as a methodological criterion is evident in Hennig's "auxiliary principle," which states that origin by convergence should not be assumed a priori (see Hennig, 1966:121). An alternative formulation might be that a single origin for similar structures and behaviors in different organisms should be presumed in the absence of evidence to the contrary. As noted by Farris (1983:12):

> Hennig's defense of the synapomorphy principle by recourse to parsimony [via his auxiliary principle] is not accidental but necessary . . . Superficially, the use of the synapomorphy principle in phylogenetic inference seems to be just a consequence of the logical connection between true synapomorphies and genealogies, but it cannot be just that, as the connection of that logic — that the traits are indeed synapomorphies — need not be met. Grouping by synapomorphy is instead a consequence of the parsimony criterion.

Hennig addressed the problems of disagreement among characters through the "method of checking, correcting, and rechecking," in an attempt to maximize congruence among character distributions. This approach has been labeled "circular" by some, because they say the process allows for the possibility of bringing observations into agreement simply for the sake of doing so, with the ultimate aim of salvaging a favored theory.

Hennig's approach has merit for the same reasons that it has been criticized. First, there is no reason to believe that initial observations are always accurate and not in need of rechecking and revalidation. This point of view derives from the idea that all observation is theory laden, and in the case of systematic character data — the theories of homology and transformation discussed in Chapters 4 and 5 — may be influenced by prior observations. Second, it is only the agreement among various data that corroborates more general hypotheses, and rechecking observations can be justified as a "corroboration" of observation as opposed to manipulation of data.

Characters or character data have frequently been referred to in the literature as "good" or "strong," on what can only be regarded as subjective grounds. For example, structurally complex characters have been deemed more reliable indicators of relationships than less complex characters on the assumption that it is hard to imagine them evolving more than once. Conversely, structurally "simple" characters have at times been relegated to a lesser status simply because they lack complexity of form. Functional significance has also played a role in judging the value of characters in some phylogenetic analyses. And, certain character systems have attracted the attention of some authors because these authors were trained in the techniques that reveal them, and they have therefore judged them more important than other types of characters. The continued development of the logic of systematic theory has brought into question all of these arguments because of their necessarily subjective quality. Each argument places a value on information, a priori, rather than examining its agreement with other data.

The strength of characters in systematics is judged by two factors: consistency and congruence.

- *Consistency* refers to the degree with which a character can be parsimoniously arranged on a given phylogenetic topology. A frequently used measure of this

Sidebar 3
Dichotomy as Part of the Method

Hennig's methods, in their original exposition, strongly implied the desirability of arriving at completely dichotomous phylogenetic schemes. This idea was almost immediately construed to mean that the methods of phylogenetic systematics only worked if the speciation process were actually dichotomous. Yet, the actual reason we desire dichotomous schemes is because they imply more informative schemes of relationships. Thus, the idea of dichotomous branching says nothing about the reality of evolution, but only how to get the most information out of our data. Furthermore, the nodes in cladograms do not represent speciation events, but rather groupings based on character information.

Hennig (1966, 209) stated with regard to dichotomy that,

Strange to say, the view is often advanced that phylogenetic systematics presupposes a dichotomous structure of the phylogenetic tree. Because dichotomy is not the rule, it is said that a system that gives the impression of a continuously dichotomous differentiation cannot be regarded as a true presentation of actual kinship relations. If phylogenetic systematics starts out from a dichotomous differentiation of the phylogenetic tree, this is primarily no more than a methodological principle. . . . *A priori* it is very improbable that a stem [ancestral] species actually disintegrates into several daughter [descendant] species at once, but here phylogenetic systematics is up against the limits of solubility of problems.

Many applications of Hennig's methods have produced schemes that were not completely dichotomous. The terms *trichotomy* (three branches) and *polytomy* (more than three branches) have become widely used when referring to such multiple branchings in cladograms.

property is the *consistency index,* which is the ratio of observed changes relative to the minimum number of changes actually required for the character(s) (see Chapter 6 for further discussion).
- *Congruence* refers to the degree with which distributions of individual characters agree with one another. This property has sometimes been described using terms such as "concordance" or "consilience." The strength of a given theory (in this case a character) derives from the fact that it is supported through repeated observation and conformity with observations on other characters.

These approaches to judging character data represent a direct application of the parsimony criterion.

The Popperian Approach to Systematics

Criticisms

Popper constructed much of his discussion of scientific methods as if it applied to the physical and experimental sciences although Popper's criticisms of positivism did not derive from an examination of the physical sciences. Substantial commentary exists in the literature suggesting the Popperian view of science requires "law-like" qualities of "process" theories (Hull, 1983:178) and cannot therefore be applied to systematic biology. This view can be shown to rest on a mistaken understanding of the nature of theory and test in systematic biology, as we will discuss below under "The Basis of Systematic Knowledge." If the purpose of Popper's writings was to distinguish between science and non-science, as he himself stated, then to conclude that his methods are inapplicable to systematics would be to conclude that systematics was not scientific or that there is more than one kind of science.

The Popperian approach to systematics has also been criticized as representing absolute or naive falsificationism. Such labels suggest that the falsificationist approach fails to recognize that observation itself is subject to error and interpretation or, more specifically, that the variability inherent in most biological systems is not taken into account by those who apply Popper's methods. Under the absolute interpretation of falsification, a theory for which any contrary evidence is adduced would be absolutely and forever discarded without consideration of the preponderance of evidence.

As an example we might consider the statement "all pterygote insects have wings." Finding a contrary example under the absolute falsificationist perspective would reject the theory even though many other attributes clearly suggest that some nonwinged organisms are pterygote insects. Such an extreme approach would not attempt to minimize the number of ad hoc statements necessary to salvage our best available theory, but rather reject it because of a single nonconforming observation. This "absolute falsificationist" criticism seems as

much a label as an argument and has been advanced by those intent on rejecting the deductivist approach, by whatever means.

Neither type of criticism seems relevant to the factual basis for systematics, nor
to what has actually taken place in the application of the Popperian approach.
One will, on examination of the literature, find numerous applications of the deductive approach in systematics, but few if any applications of absolute falsificationism. We might identify, however, no fewer than three issues important to the
positive development of systematics as a science that at least in part derive from
critical attempts to apply the deductive approach.

Benefits

First, whereas previously it was at best unclear what represented hypotheses
to be judged (tested) in systematics, it is now abundantly clear that systematic
hypotheses are characters and the hierarchic schemes of relationships they imply. Characters represent lower-level theories closely associated with the realm
of observation and the formulation of ideas concerning homology. Hierarchic
schemes of relationships represent higher-level theories that are tested with the
lower-level character theories. Failure of agreement between theory and observation, once explained away with ad hoc hypotheses, is now adjudicated via the
parsimony criterion in the search for theories of relationships among taxa that
show greatest agreement with available data.

Second, the meaning of prediction as applied to systematics has been clarified. This issue was previously muddled, notwithstanding its central role in characterizing the "power" of science. For example, Bock (1974) unequivocally related classical evolutionary taxonomy to Popper's philosophy, but at the time was
decidedly unclear on the issues of prediction and testing. More recently, Mayr
(1988:19) seemed equally unclear when he stated that a belief in universal laws
implies a belief in absolute prediction and that because of the pluralism of cau-
sations and solutions in biology, prediction is probabilistic, if possible at all.

In spite of the views espoused by Bock and Mayr, prediction in systematics is
now widely — if not universally — understood to mean the ability to predict the
distributions of previously unobserved (unstudied) characters among taxa. The
strength of phylogenetic hypotheses is based not only on what predictions about
character distributions can be made but also on what character distributions are
prohibited. Lest there be further confusion, it must be emphasized that prediction as applied in systematics does *not* refer to the forecasting of future evolutionary events, as has sometimes been suggested.

Third, the issue of how we judge the explanatory power of systematic theories
through the application of parsimony has been clarified. Ad hoc hypotheses
are minimized; congruence among character distributions is maximized. Thus,
agreement between theory and observation is maximized.

What seems largely irrelevant in evaluating the deductivist approach is whether or not all of the elements of current "evolutionary thinking" were embraced, whether the approach of Popper represents "absolute" falsificationism, and whether theories in systematics — and possibly biology more generally — can be of metaphysical origin. Clearly, advancements in the science of systematics have been achieved by applying the Popperian approach within the confines of biological reality, in spite of all arguments to the contrary.

Engelmann and Wiley (1977) and Gaffney (1979a, 1979b) emphasized the significance of the falsifiability criterion in systematics, and particularly with relation to paleontology. They noted that ancestor–descendant relationships, as commonly invoked in the paleontological literature, are not testable in the Popperian sense and that such "schemes" can only be labeled as scenarios. Gaffney (1979b) concluded — with regard to scenarios — that even though it would be wonderful to have testable hypotheses of selection pressure and adaptive zones for Devonian tetrapod vertebrates, we must construct classifications through the formulation of hypotheses testable with available information. Gaffney saw no data available for testing theories concerning selection pressures in the Devonian (see further discussion in Chapter 4 under Ancestors, Sister Groups, and Age of Origin).

Statistics, Probability, and Models as Alternatives to Parsimony

The application of parsimony in judging phylogenetic theories has been criticized on the grounds that evolution is not necessarily parsimonious. This argument suggests that we must know something about the nature of the products of evolution before we conduct the studies designed to reveal that knowledge. As with all scientific inquiry, parsimony is applied in systematics to minimize the number of ad hoc explanations of data or, stated in another way, to maximize the power of the hypothesis to explain those data. The current sense of its application is in no way related to models of evolution. Parsimony has to do with the interpretation of evidence. It is a methodological tool and makes no claims about the way things are in the world.

On the above view, parsimony might be said to be free of assumptions concerning evolutionary models. But this perspective is not unchallenged in the literature. For example, Sober (1983) stated that he believed parsimony contained presuppositions, and although he was unclear as to what they might be, it was his view that if the history of evolution violates them, then character distributions will not be usable in reconstructing phylogenies.

Swofford et al. (1996:426), in a hermeneutical discussion, asserted that parsimony does make assumptions and that the violation of these assumptions can lead to problems. In their view, "The difficulty lies in stating precisely what the assumptions are." This statement apparently means that just because no authors have explicitly stated a set of assumptions for parsimony as a methodological criterion, there is no reason to suppose that such assumptions do not exist or that they will not eventually be discovered and stated. As a way of providing a boundary for what they believed to be the yet unstated conditions implied in the application of parsimony, Swofford et al. went on to say that "At a minimum, acceptance of an optimal tree under the parsimony criterion requires one to assume that conditions that can cause parsimony to estimate an incorrect tree are unlikely to have occurred."

These statements might be interpreted as suggesting that no criterion exists for judging the results of parsimony analyses, whereas in reality the criterion is just that the data are explained with minimal unnecessary interpretation. There are no confidence limits to be calculated for judging the truth content of a given hypothesis under the parsimony criterion. The standard of judgment is comparison with alternative hypotheses.

A final word on assumptions might best be made with a quote from the work of Farris (1983:35):

> Parsimony analysis is realistic, not because it makes just the right suppositions on the course of evolution. Rather, it consists exactly of avoiding uncorroborated suppositions whenever possible. To the devotee of supposition, to be sure, parsimony seems to presume very much indeed: that evolution is not reversible, that rates of evolution are not constant, that all characters do not evolve according to identical stochastic processes, that one conclusion of homoplasy does not imply others. But parsimony does not suppose in advance that those possibilities are false — only that they are not already established. The use of parsimony depends just on the view that the truth of those — and any other — theories of evolution is an open question, subject to empirical investigation.

The viewpoint of Swofford et al. quoted above is associated with an argument originally propounded by Felsenstein (1978), who concluded that because parsimony did not provide the known "true" answer in simulations based on certain models of character evolution, parsimony must therefore be flawed because it was statistically "inconsistent." Felsenstein reasoned from this result that phylogenetic theories might better be judged on a statistical basis, through the use of a maximum-likelihood approach (see further discussion of Maximum Likelihood and Long-Branch Attraction in Chapter 6).

Farris (1983:17), in criticizing the application of a statistical approach such as maximum likelihood to phylogenetic inference, noted that it

> was wrong from the start, for it rests on the idea that to study phylogeny at all, one must first know in great detail how evolution has proceeded. That cannot very well be the way in which scientific knowledge is obtained. What we know of evolution must have been obtained by other means. Those means . . . can be no other than that

phylogenetic theories are chosen, just as any scientific theory is, for their ability to explain available observations.

Felsenstein (1978:409) himself admitted that "The weakness of the maximum likelihood approach is that it requires us to have a probabilistic model of character evolution which we can believe. The uncertainties of interpretation of characters in systematics is so great that this will hardly ever be the case." Yet he asserted that the only true tests of the robustness of phylogenetic methods must be undertaken statistically and that this goal must be pursued even though "Establishing that robustness . . . by examining a wider range of models is a daunting task."

Parsimony was further impugned by Felsenstein (1981:194) because it "implicitly assume[s] very low rates of change." Farris (1983:13) discussed the issue of whether a "procedure that minimizes something must ipso facto presuppose that the quantity minimized is rare." He used an example from regression analysis, pointing out that in statistics, just because the variance may be large, there is no assumption that the regression line found is not the best description of the data, just that the confidence limits must be expanded to accommodate the variance.

The statistical worldview would seem to suggest that we can construct some absolute measure of the goodness of phylogenetic hypotheses. This view of science, when applied as a criticism of parsimony, disregards the fact that parsimony procedures have always been applied as a way of judging the explanatory power of *alternative* hypotheses, not as a way of understanding the truthfulness of hypotheses in absolute or probabilistic terms.

The Basis of Systematic Knowledge

In justifying his arguments establishing a general reference system for biology on the basis of phylogeny, Hennig concluded that in a system of descent with modification, once modifications are fixed in lineages, we can expect to observe a hierarchic distribution of attributes (Hennig, 1966; Davis and Nixon, 1992). The natural hierarchy of attributes of organisms, then, is an operational assumption of the method for discovering relationships (Brady, 1983). It is the fixity of attributes at the minimum level, what we will call *species,* that admits the units as valid for analysis in such a system (Davis and Nixon, 1992).

Hennig's view of the goals of systematics was heavily grounded in biology. A more philosophically oriented statement of purpose was offered by Hull (1965, in Ereshefsky, 1992:202), who declared that "From the beginning, taxonomists have sought two things — a definition of 'species' which would result in real species and a unifying principle which would result in a natural classification." Hull's statement points to two lines of argument — which overlap at times. The issues raised by Hull boil down to whether systematics informs our knowledge of evolution, and particularly evolutionary history, or whether knowledge of evolution is a prerequisite to understanding systematic relationships.

It has been argued that taxonomy has a history tainted with "essentialism," and that only when the undesirable influences of that approach are removed will the scientific status of the field be established. This is the "species problem" of Hull. Hull (1965) asserted that taxon names (including species) cannot be defined by sets of properties because to do so would require that the properties be distributed both universally and exclusively among the members of a taxon. Otherwise, the definitions would not be universally applicable. This view was adopted by de Queiroz and Gauthier (1990), who noted that numerous taxonomic papers contain a section entitled 'definition,' which indicates that the names of taxa are being defined by lists of organismal traits and that such 'definitions' must therefore be directly associated with the definition of an essence in the Aristotelian sense. In the view of de Queiroz and Gauthier, definition stipulates some necessary and sufficient conditions, and the presumably invariant nature of the properties on which the definition is based. What seems apparent, however, is that "define" as used in most papers was never intended in any such sense.

The purportedly pernicious influence of essentialism on systematics is found in the works of Mayr (1982) under the heading of "typological thinking." Mayr's worldview emphasizes "population thinking" and the importance of variation in biological systems in general and in "species" in particular. Mayr has labeled most approaches that are not heavily based on the analysis of variation at the species level and below as typological or essentialist. These categorizations would seemingly include most pre-Darwinians as well as a large proportion of more modern workers. Hull, Mayr, de Queiroz and Gauthier, and Ghiselin (1984) would have us believe that essentialism, in the sense that they understand it, is actually being practiced by taxonomists and that it must be eradicated to set things right in systematics as a science.

As noted by Nelson and Platnick (1981:328), the anti-essentialist argument assumes that when the attributes of a lineage change over time, the modified versions of those attributes will lose their genealogical connections and be unrecognizable. For example, Nelson and Platnick observed that under this presumptive view of taxonomy, we could not use characters such as fins as part of the definition of the group Vertebrata because some members of that group, the tetrapods, may possess fins in a modified form, as limbs. Nonetheless, as we will see, the concept of character transformation plays a central role in the systematic approach of even the anti-essentialists, and all systematists make use of that concept in forming hierarchic classifications.

Hull's (1965) second imputed aspiration of taxonomists, the "unifying principle which would result in a natural classification," also requires scrutiny in our efforts to understand the basis of systematic knowledge. The 'unifying principle' to which Hull alluded is widely thought to be the theory of organic evolution. Descent with modification is certainly capable of producing the hierarchic patterns of relationships that taxonomists have observed "from the beginning" in the jargon of Hull. However, descent with modification, in its status as a unifying principle does not provide — by itself — a method for discovering that hierarchy. Hull, the philosopher, having written his tract in 1965, might be forgiven for lack of foresight. Forgiven, because the details of a method that is capable of effectively and consistently recovering information on the hierarchy of life was just at the point of becoming widely appreciated in 1965. What seems less clear is why subsequent authors — particularly biologists — have adopted Hull's anti-essentialist arguments lock, stock, and barrel long after the distinction between a theory capable of explaining hierarchy (i.e., organic evolution) and a method capable of discovering the hierarchy of life (cladistics) had been clarified.

Hull's argument concerning natural classifications has been used against practitioners of cladistic methods — including Gareth Nelson, Norman Platnick, and others — who have argued forcefully that cladistics is a system of pattern recognition and that a priori assumptions about evolution are unnecessary for the success of the method (Platnick, 1985). Beatty (1982) derided what he called "pattern cladistics" because the approach did not explicitly justify the search for a hierarchic pattern of relationships on what he viewed as accepted tenets of biological evolution. Beatty's criticisms dismiss the idea that the theory of descent with modification gains credence through an independent source of evidence provided from the recovery by systematists of hierarchic patterns of relationships among taxa.

In responding to the arguments of Beatty, Hull, and others, Platnick (1982) wondered what causal theory was necessary to recognize the group known as spiders comprising 35,000 species of animals, all of which are united by their possession of abdominal spinnerets and male pedipalps modified for sperm transfer. The same might be said for the 1 million species of winged insects, 9,000 species of winged and feathered birds, 250,000 species of flowering plants, and many other groups, all of which were recognized long before the formulation of a coherent theory of organic evolution. Our ability to recognize such groups in nature is based on the structural attributes that each possesses and the congruence of distributions among those attributes. The relationships among the groups are established on the basis of structural *similarities,* not on structural *identities.* Explaining the existence of hierarchically related systematic groupings is facilitated by an overarching explanatory theory, in this case, organic evolution. However, as in other branches of science, our ability to make such discoveries does not depend on such an explanatory theory.

An analogy of planetary motion was invoked by Platnick (1982) as a way of bringing this issue into perspective. He asked if a causal theory was required to discover the existence and movements of Jupiter. The answer is no. We might extend Platnick's line of argumentation by observing that the ancient Greeks, Semites, and Chinese had detailed knowledge of planetary existence, even in the absence of telescopes, and were capable of predicting the positions of celestial bodies with considerable accuracy, in part evidenced today by use of calendars (and horoscopes) derived from these cultures. This detailed knowledge of the heavens was developed in spite of the fact that the earth was thought to be the center of the universe.

Platnick (1982) reckoned that authors such as Beatty, who argue for the primacy of evolutionary theorizing over systematics, have the cart before the horse, both historically and logically. In different words, Patterson (1987: 4) stated the case thus: "In the decades after the [appearance of] *The Origin of Species,* comparative morphology became phylogenetic research; common ancestors replaced archetypes, and homology became evidence of common ancestry rather than of common plan [essence]. But these were changes in doctrine rather than of practice."

Perhaps the most elegantly developed argument for the "independence" of systematics is that of Brady (1985:114), who argued that

> the hierarchy of taxa was considered by Darwin [in *The Origin of Species*] to be one of the established facts of natural history ... The hierarchy and other patterns of natural history were not, therefore, based upon evolutionary theory [in the mind of] the founder of that theory. They occupied the privileged position of an independent evidence, to which one could point for a test of the theory [of evolution].

Brady further observed that systematics is complete in itself, generating its own standards and tests, and that the addition of evolutionary theory comes after the discovery of the patterns. In Brady's view (p. 125) it seemed clear that most cladists discovered the patterns of relationship by the same method, and he therefore found it odd that those singled out for attack as "pattern cladists" were the ones who insisted that what we know about the world are the patterns derived from observation rather than the explanations for those patterns. In Brady's view, it is the independence of those patterns that allows systematics to retain its empirical status. Failure to distinguish the empirical problem (the pattern of relationships among taxa) from the explanatory hypothesis (the theory of organic evolution) leaves us with no independent evidence to test that hypothesis. As stated by Brady (1985:117), "by making our explanation into the definition of the condition to be explained, we express not scientific hypothesis but belief. We are so convinced that our explanation is true that we no longer see any need to distinguish it from the situation we are trying to explain."

For the moment, let us summarize by saying that *systematics* is the branch of biology capable of providing a genealogical history of life on earth. The methods advocated might or might not reveal the truth, but they do allow us to interpret the history of relationships among organisms in a way that conforms as closely as possible to the data at hand. *Cladistics* offers a logically coherent framework integrating observation, analysis, and synthesis of systematic information.

The most general conclusion to be derived from this discussion is that patterns of phylogenetic relationships are neither directly observable nor directly knowable. Rather, acquisition of such knowledge comes from the collection of relevant data and the analysis of those data by appropriate methods. The methods of data acquisition and analysis are described in the following chapters.

Literature Cited

Beatty, J. 1982. Classes and cladists. *Syst. Zool.* 31:25–34.

Bock, W. 1974. Philosophical foundations of classical evolutionary classification. *Syst. Zool.* 22:375–392.

Brady, R. H. 1983. Parsimony, hierarchy, and biological implications. pp. 49–60. *In:* Platnick, N. I., and V. A. Funk (eds.), *Advances in Cladistics,* Vol. 2. Proceedings of the Second Meeting of the Willi Hennig Society. Columbia University Press, New York.

Brady, R. H. 1985. On the independence of systematics. *Cladistics* 1:113–126.

Davis, J. I., and K. C. Nixon, 1992. Populations, genetic variation, and the delimitation of phylogenetic species. *Syst. Biol.* 41:421–435.

de Queiroz, K., and J. Gauthier. 1990. Phylogeny as a central principle in taxonomy: phylogenetic definitions of taxon names. *Syst. Zool.* 39:307–322.

Dingus, L., and T. Rowe. 1998. *The Mistaken Extinction: Dinosaur Evolution and the Origin of Birds.* W. H. Freeman and Co., New York. 332 pp.

Engelmann, G. F., and E. O. Wiley. 1977. The place of ancestor-descendant relationships in phylogeny reconstruction. *Syst. Zool.* 26:1–11.

Farris, J. S. 1983. The logical basis of phylogenetic analysis. pp. 1–36. *In:* Platnick, N. I., and V. A. Funk (eds.), *Advances in Cladistics,* Vol. 2. Proceedings of the Second Meeting of the Willi Hennig Society. Columbia University Press, New York.

Felsenstein, J. 1978. Cases in which parsimony or compatibility methods will be positively misleading. *Syst. Zool.* 27:401–410.

Felsenstein, J. 1981. A likelihood approach to character weighting and what it tells us about parsimony and compatability. *Biol. J. Linn. Soc.* 16:183–196.

Gaffney, E. S. 1979a. An introduction to the logic of phylogeny reconstruction. pp. 79–111. *In:* Cracraft, J., and N. Eldredge (eds.), *Phylogenetic Analysis and Paleontology.* Columbia University Press, New York.

Gaffney, E. S. 1979b. Tetrapod monophyly: a phylogenetic analysis. *Bull. Carnegie Mus. Nat. Hist.* 13:92–105.

Ghiselin, M. T. 1984. "Definition," "character," and other equivocal terms. *Syst. Zool.* 33:104–110.

Hennig, W. 1966. *Phylogenetic Systematics.* University of Illinois Press, Urbana. 263 pp.

Hull, D. L. 1965. The effect of Essentialism on taxonomy: Two thousand years of stasis. *Brit. J. Phil. Sci.* 15:314–326; 16:1–18 (reprinted in: Ereshefsky, M. [ed.]. 1992. *The Units of Evolution: Essays on the Nature of Species.* MIT Press, Cambridge, Massachusetts. 405 pp.).

Hull, D. L. 1983. Karl Popper and Plato's metaphor. pp. 177–189. *In:* Platnick, N. I., and V. A. Funk, (eds.), *Advances in Cladistics.* Vol. 2. Proceedings of the Second Meeting of the Willi Hennig Society. Columbia University Press, New York.

Mayr, E. 1982. *The Growth of Biological Thought: Diversity, Evolution, and Inheritance.* Harvard University Press, Cambridge, Massachusetts.

Mayr, E. 1988. *Toward a New Philosophy of Biology: Observations of an Evolutionist.* Harvard University Press, Cambridge, Massachusetts. 564 pp.

Nelson, G., and N. Platnick. 1981. *Systematics and Biogeography: Cladistics and Vicariance.* Columbia University Press, New York. 567 pp.

Patterson, C. 1987. Introduction, pp. 1–22. *In:* Patterson, C. (ed.), *Molecules and Morphology in Evolution. Conflict or Compromise?* Cambridge University Press, Cambridge.

Platnick, N. I. 1979. Philosophy and the transformation of cladistics. *Syst. Zool.* 28:537–546.

Platnick, N. I. 1982. Defining characters and evolutionary groups. *Syst. Zool.* 31:282–284.

Platnick, N. I. 1985. Philosophy and the transformation of cladistics revisited. *Cladistics* 1:87–94.

Popper, K. 1968. *The Logic of Scientific Discovery.* Second English edition. Harper and Row, New York.

Popper, K. 1975. The rationality of scientific revolutions. pp. 72–101. *In:* Harre, R. (ed.), *Problems of Scientific Revolution: Progress and Obstacles to Progress in Science,* The Herbert Spencer Lectures, 1973. Clarendon Press, Oxford.

Rieppel, O. C. 1988. *Fundamentals of Comparative Biology.* Birkhauser, Basel. 202 pp.

Sober, E. R. 1983. Parsimony methods in systematics. pp. 37–47. *In:* Platnick, N. I., and V. A. Funk (eds.), *Advances in Cladistics,* Vol. 2. Proceedings of the Second Meeting of the Willi Hennig Society. Columbia University Press, New York.

Swofford, D. L., G. J. Olsen, P. J. Waddell, and D. M. Hillis. 1996. Phylogenetic Inference. 407–514. *In:* Hillis, D. M., C. Moritz, and B. K. Mable (eds.), *Molecular Systematics,* second edition. Sinauer Associates, Sunderland, Massachusetts.

Suggested Readings

Brady, R. H. 1985. On the independence of systematics. *Cladistics* 1:113–126.

Gaffney, E. S. 1979. An introduction to the logic of phylogeny reconstruction. pp. 79–111. *In:* Cracraft, J., and N. Eldredge (eds.), *Phylogenetic Analysis and Paleontology.* Columbia University Press, New York. [A clear and concise discussion of the hypothetico-deductive approach to systematics]

Magee, B. 1973. Karl Popper. pp. 10–49. *In:* Kermode, F. (ed.), *Modern Masters.* Viking Press, New York. [An easily understood synthesis of the philosophy of science propounded by Karl Popper]

Popper, K. 1968. *The Logic of Scientific Discovery.* Second English edition. Harper and Row, New York.

Rieppel, O. C. 1988. *Fundamentals of Comparative Biology.* Birkhauser, Basel. 202 pp.

Siddall, M. E., and A. G. Kluge. 1997. Probabilism and phylogenetic inference. *Cladistics* 13:313–336.

Wiley, E. O. 1975. Karl Popper, systematics, and classification: A reply to Walter Bock and other evolutionary systematists. *Syst. Zool.* 24:233–243.

SECTION II

CLADISTIC METHODS

4

Homology and Rooting

Having laid the groundwork for a general understanding of systematics and the philosophy of science, we will now begin the detailed examination of methods for establishing hierarchic — phylogenetic — relationships among taxa. The concepts of synapomorphy and monophyly were central to the method of phylogenetic analysis developed by Hennig and were unequivocally characterized by him. Along with homology and rooting by the outgroup method, they form the focus of this chapter.

Homology and Synapomorphy

Biologists have produced substantial evidence over the last two centuries for the mechanisms of inheritance and for ontogenetic transformation. Yet, there is little or no direct evidence for many of the grand-scale transformations that we see in phylogeny: the development of the true arthropod body plan, irrespective of its similarity to that of annelids and onychophorans; the development of insect wings; the development of turtles from more generalized tetrapods; the origin of flowers in the angiosperms. It had long been hoped that paleontology would provide the evidence in the form of intermediate fossils, the missing links. Although evidence exists for the development of tetrapods from fishlike ancestors, for most groups of organisms there are no intermediate fossils and consequently no direct evidence of transformation. What we have, by way of explanation, is the concept of "punctuated equilibrium" (Eldredge and Gould, 1972). Yet, this does not *explain* transformation, but only suggests that it has apparently occurred relatively rapidly, at intervals, between long periods of comparative morphological stasis.

Natural historians have observed structural correspondences among organisms since the time of the ancient Greeks. These similarities of form were termed homologous by the British anatomist Richard Owen (e.g., Patterson, 1982). Owen (1843) defined *homology* as pertaining to "the same organ . . . under every variety of form and function." The criteria used by Owen were the "principle of connections" and the "principle of composition," both of which, according to Brady (1985) and Rieppel (1988), appeared earlier in the work of Geoffroy St. Hilaire (Geoffroy, 1818). Theories of homology in the sense of Geoffroy and Owen exist independent of the theory of evolution. The "common plan" of homologies found in nature constitutes the empirical pattern of transformation that allows us to postulate connections among groups of organisms. It is only through theories of homology that phylogenetic analysis can proceed. The minimal assumption of the homology concept in cladistic analysis seems to be that features of organisms can be — or are — transformed over time.

The modern formulation of the criteria for recognizing homology are frequently attributed to Remane (1952), and can be stated as follows:

1. position (similarity of topographical relationships)
2. similarity of special structure
3. connection by intermediates (transformation)

Mayr (1982:233) complained that it "was unfortunate, and quite inappropriate, that Remane raised the criteria that served as *evidence* for homology to the *definition* for homology." According to Mayr (1982:232) "After 1859 there has been only one definition of homologous that makes biological sense: A feature . . . is homologous in two or more taxa if it can be traced back to . . . the same . . . feature in the presumptive common ancestor of these taxa." Brady (1985:116–117) criticized Mayr's reasoning by noting that it confused the condition to be explained (similarity of structure) with the explanation itself (theory of organic evolution). In Brady's view, Remane was thinking clearly when he formulated his criteria for the recognition of homology.

Considering Hennig's viewpoint with regard to homology may help clarify the position of Brady and the one taken in this book, while at the same time revealing the fallacy of Mayr's argument. Hennig (1966:93–94) observed that "Apparently it is often forgotten that the impossibility of determining directly the essential criterion of homologous characters — their phylogenetic derivation from one and the same previous condition — is meaningless for defining the concept of homology." To make his point, Hennig quoted Boyden (1947): "Today the pendulum has swung so far from the original implication in homology that some recommend that we define homology as any similarity due to common ancestry, as though we could know the ancestry independently of the analysis of similarities!"

Observations on similarity of structure — homologies — across a set of taxa lead to three possible conclusions, which are illustrated graphically in Fig. 4.1:

• The features thought to be homologous are invariant among the taxa being examined, and neither support nor contradict any groupings of them.
• The features thought to be homologous vary in such a way as to form nested groupings of the taxa and to define those groupings in an unequivocal manner.

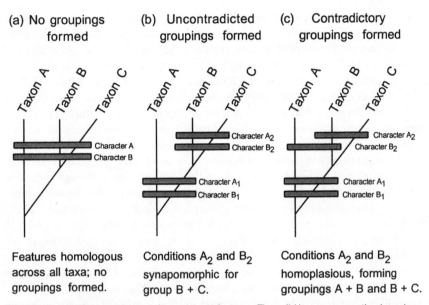

(a) No groupings formed

(b) Uncontradicted groupings formed

(c) Contradictory groupings formed

Taxon A   Taxon B   Taxon C

Character A
Character B

Character A_2
Character B_2

Character A_1
Character B_1

Character A_2
Character B_2

Character A_1
Character B_1

Features homologous across all taxa; no groupings formed.

Conditions A_2 and B_2 synapomorphic for group B + C.

Conditions A_2 and B_2 homoplasious, forming groupings A + B and B + C.

Fig. 4.1. The possible distribution of homologous features. The solid bars, representing homologs, connect the groups in which they occur. The numbered subscripts indicate the varying conditions for each homologous feature.

The conditions of homologous features that uniquely define groupings at different levels are *synapomorphies*.

• The features thought to be homologous define different and conflicting groupings. Features that are contradictory with regard to their group-defining ability are *homoplasies*.

In the first instance, the theory of homology remains uncontradicted, but no further conclusions can be drawn. In the second case, the theory of homology is valid, but two properties are seen in addition, those being the nested nature of the transformed properties and the fact that nested transformations in two homologous features define the same hierarchy. The third case presents contradictions that might cause us to question our theories of homology based on observation or to conclude that there is more than one possible hierarchy of relations.

We can see from this example that homologous structures can exist in a more general condition, whereby in Fig. 4.1, conditions A_1 and B_1 are *plesiomorphic* relative to conditions A_2 and B_2, respectively. Or, they may exist in a more restricted condition, whereby conditions A_2 and B_2 are *apomorphic* relative to conditions A_1 and B_1, respectively. It is the apomorphic condition that is important in the recognition of monophyletic (natural) groups. Characters are usually termed *synapomorphic* when occurring in more than one taxon, and *autapomorphic* when occurring in a single taxon. The use of the terms is therefore relative.

Hennig argued that *synapomorphies* — shared derived characters — provide the only evidence for the existence of natural groups. This is the singular aspect of his arguments for the phylogenetic system. All of Hennig's other principles are subsidiary to it. Thus, synapomorphy and homology are directly related. Nearly all formulations of systematic methods have treated the recognition of "homologies" as central to the success of the method. Some of those formulations have failed, however, because they did not recognize that even though all members of a group may share some homologous feature in common, that feature could not be group-defining if it also existed outside the group. Possibly the most prominent examples of characters that also exist outside the groups they are used to define are poikilothermy and scales in the Reptilia — even though these occur in most bony fishes — and two cotyledons in the dicotyledonous angiosperms — which also occur in the Cycadales, Ginkgoales, Pinales, and Gnetales. Thus, homologies — as synapomorphies — form a nested set of relationships, and it is only when the hierarchic relationships of groups conform to the nesting of synapomorphies that a "natural" classification results.

To further clarify the issue, let us use the example of tetrapod forelimbs (Fig. 4.2). The limbs of terrestrial vertebrates represent novel attributes — synapomorphies — relative to the pelvic and pectoral fins of fishes. These limbs do not represent distinctive attributes for bats, birds, and pterosaurs. Yet, wings in each of those groups are distinctive — apomorphic — for them. The concept of transformation is exemplified by the modification of pectoral fins into forelimbs and forelimbs into wings, the latter on at least three independent occasions, for birds, bats, and pterosaurs. Whereas the vertebrate pectoral fin–forelimb in all of its forms might be referred to as homologous, it is the nesting of the structural types into which it has been transformed that allows us to postulate a hierarchy of relationships. The structural types represent synapomorphies for the groups that possess them. At the most general level, fins are synapomorphies. At a more restricted level, limbs are synapomorphic. And at the most restricted levels, wings are apomorphic, for three separate groups. The postulated transformation is graphically depicted in Fig. 4.2. Additional examples are seen in Fig. 4.3.

Tests of Homology

Homology in the sense used above is strictly associated with the topographical and structural relations of similarity. Theories of homology based on observation in a single taxon may be tested in two ways when those observations are made across multiple taxa: conjunction and similarity of structure and position.

Conjunction. The conjunction test dictates that multiple homologs may not exist in the same organism. Under this criterion the apparent multiple occurrence of an organ or structure in a single organism would indicate that nonhomolo-

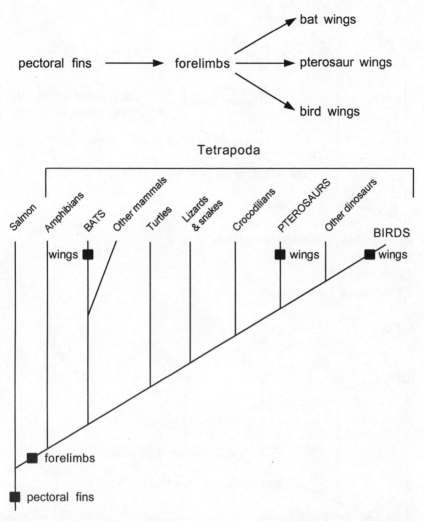

Fig. 4.2. Nesting of structural transformation of the vertebrate forelimb. Note that the modification of fins into limbs represents a synapomorphic character for the tetrapod vertebrates. The three distinctive modifications of forelimbs into wings represent apomorphic characters for restricted groupings within the Tetrapoda.

gous structures have been mistakenly treated as the same. The conjunction criterion serves primarily to exclude false concepts of homology. There are, however, situations that might be considered exceptions to the conjunction test. First is *serial homology,* as for example the segmental repetition of body parts in the Annelida and Arthropoda. Second is *homonymy,* sometimes called mass homology. Prominent examples include the occurrence of multiple leaves on a single plant and hair on mammals.

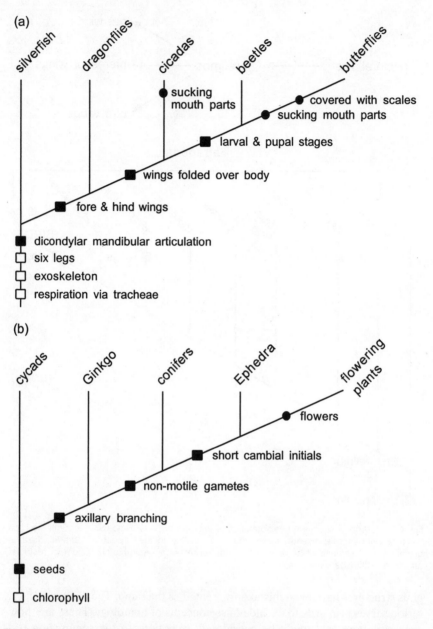

Fig. 4.3. The distribution of plesiomorphic and apomorphic characters in (a) insects and (b) seed plants. Open boxes = plesiomorphic. Closed boxes = synapomorphic. Closed circles = autapomorphic.

Similarity of Structure and Position. This is the classic test of homology: when features of similar structure and position are found in two or more different organisms, they are judged to be homologous. The connection by intermediate forms (e.g., limbs) often allows for the postulation of theories of structural homology between less similar-appearing features (e.g., fins and wings). Rejection of a theory of homology based on these criteria might involve observations of structures in additional organisms, which point to errors in the original postulation of homology on the basis of topographical position and structural similarity.

Consider the tetrapod limb as it exists in frogs, salamanders, mammals, and "reptiles." There would seem to be little disagreement in the scientific literature, or on the part of the well-informed nonscientific public, that the forelimbs of all of these groups represent the same structures — that they are homologous — as are the hind limbs among all organisms that possess them. We might conclude that these judgments are based on the observations that the limbs occupy the same positions on the body, have the same bony elements, and in most cases have the same number of digits. The limbs are, to be sure, of similar position and structure.

Within these groupings we might further consider the wings of birds and bats. Few would argue against the idea that these structures represent the forelimbs in a modified form. Such a conclusion, in the minds of most observers, would be based on similarity of position and structure, even though the bony elements have been modified, sometimes to the point of loss, as in the digits of birds.

Taking our comparisons one step further to snakes (and some other terrestrial vertebrates), we may find little or no evidence of limbs, fore or hind. In those cases we have two choices: assume that the organisms never had limbs or assume that the transformation has been toward complete loss of all structures pertaining to the limbs. The absence of limbs offers no evidence on the nature of limbs themselves — except that they are absent. Corroboration of the primitive absence theory or the loss theory cannot come directly from the structures themselves. Therefore, in such circumstances we need another basis for establishing connections among the structures, the criterion for which is discussed under "Test of Synapomorphy."

Test of Synapomorphy

Synapomorphies, then, can define groupings at different levels. Hennig referred to this phenomenon as the "heterobathmy of synapomorphy." Homologies had long been recognized, as had a systematic hierarchy, but the failure to consistently recognize the relationship between transformation and hierarchy had caused many groups to be formed on the basis of features that were not unique or on the absence of characters, thus creating unnatural groups. The observation that characters were distributed in an unorderly way in these unnatural groups was sometimes called "mosaic evolution" by Mayr and others and cited as a confounding limitation of cladistic methods (see also Farris, 1971).

This ill-defined concept of mosaic evolution offered an evolutionary explanation for the apparently disorderly origin of characters. In reality, such groups were often formed on the basis of *plesiomorphies,* attributes which were diagnostic for a group at a higher level.

Congruence. Congruence represents the test for theories of homology as synapomorphy. Congruence exists when multiple homologous features are observed as defining the same grouping of organisms. As quoted by Brady (1985:120), character congruence as a powerful force for judging the weight of evidence was already well understood by Darwin:

> We may err . . . in regard to single points of structure, but when several characters, let them be ever so trifling, occur together throughout a large group of beings having different habits, we may feel almost sure, on the theory of descent, that these characters have been inherited from a common ancestor. And we know that such correlated or aggregated characters have an especial value in classification (Darwin, 1859:426).

It is congruence that will help to resolve the question of primary absence or secondary loss of limbs in snakes. Snakes, on the basis of attributes other than limb structure, show relationships with lizards. Hypothesizing secondary loss of limbs in snakes is "congruent" with the distribution of those other attributes. Asserting that snakes have never had legs, or arose from legless ancestors, would place them outside the remaining vertebrates and would require that most of their osteology and anatomy be independently derived. Thus, in this case, congruence as understood through the application of the parsimony criterion offers the strongest evidence that the ancestor of snakes possessed limbs, even though the evidence might be construed as indirect. The same arguments could be made for the absence of chlorophyll in saprophagous angiosperms and the absence of wings in lice and fleas.

Interpreting the Literature on Homology and Synapomorphy

Most authors agree that the basis of phylogenetic analysis rests on theories of homology. However, comparing even the most thorough and cogent available discussions of homology and its bearing on phylogeny reconstruction (e.g., Patterson, 1982; Rieppel, 1988; de Pinna, 1991) may leave the reader confused. This confusion may arise because the same terms are applied to related, but different, concepts and because the nature of tests for "lower-level" (homology) theories and "higher-level" (synapomorphy) theories are also in dispute. We will discuss these two aspects in turn.

Terminology. The relationship of the terminology and concepts discussed above, as applied by various authors, is shown in Table 4.1. We can see that some au-

Table 4.1. Homology: Relationships of Concepts and Terminology

Author	Topographical & Structural Similarity	Character Congruence	Character Incongruence
Owen	homology	— — — —	analogy
Lankester	homology	homogeny	homoplasy
Patterson	homology-synapomorphy	synapomorphy	homoplasy
dePinna	primary homology	secondary homology	homoplasy
Lipscomb	homology	synapomorphy	homoplasy
Brower & Schawaroch	observation	homology-synapomorphy	homoplasy
Hennig	homology	synapomorphy	convergence
Mayr	homology	homology	convergence/parallelism
G. Ball	homology	synapotypy	convergence/parallelism

Adapted and expanded from Rieppel (1988).

thors have used *homology* in the sense it is applied in this book, to refer solely to "topographical and structural relations of similarity." Others, however, have used the term as a synonym of synapomorphy, leaving "homology as observation" without an applicable term. The term "primary homology" has been used by de Pinna (1991) as a substitute. Apomorphy would seem to be synonymous with "homogeny" of Lankester (1870). Whereas the former term, as proposed by Hennig, has now achieved widespread usage, the significance of the latter was apparently not appreciated at the time of its proposal and consequently was never widely adopted.

Tests. With regard to tests, we might first consider the viewpoint of Brower and Schawaroch (1996:268), who argued that "homology cannot be determined prior to cladistic analysis." From this perspective, repeated observation of similarity of structure across a group of taxa is not a test of homology but rather a factor that compels us to postulate homology. A contrary view was advocated by Lipscomb (1992), who considered repeated observation of similarity a valid test of homology, such that among all possible transformation series for the set of possible conditions of a homologous feature, those that postulate transformations between states that are least similar will be rejected. Lipscomb (1992:52), unlike de Pinna (1991), argued that theories of homology are based on "the meticulous examination of all details" and concluded unequivocally that such theories are available for test at the level of observation and simple comparison across taxa.

In agreement with the point originally emphasized by Patterson (1982), nearly all authors — including de Pinna, Lipscomb, and Rieppel — have concurred that synapomorphy is tested by character congruence. The differences of opinion among these and other authors have to do primarily with the varied application of terms, as seen in Table 4.1.

Homoplasy (Convergence and Parallelism)

Hennig argued that homology — in the form of observed similarity — should be accepted in the absence of evidence to the contrary. The approach was introduced in Chapter 3 under the heading of parsimony. As explained by de Pinna (1991:371):

> All similarities are deemed homologous initially, and non-homology is disclosed by a pattern-detecting procedure [parsimony] . . . If the analysis supports a single position for a putative synapomorphy, then the condition shared by the various taxa with that derived condition are [sic] corroborated as homologous. If a shared derived condition turns out to require independent origins in the overall scheme of relationships, then an event of non-homology [homoplasy] has been discovered.

Thus, as shown in Table 4.1, independent theories of homology may produce conflicting conclusions when testing theories of group relationships. The observer is then forced to conclude either that similar structures arose more than once through parallel or convergent evolution as analogs or that existing approaches do not enable us to distinguish between things that are actually different. In the latter case we are left with the conclusion that our observations are most likely mistaken. The apparent multiple evolution, reduction, or re-evolution of structures was first referred to as *homoplasy* by Lankester (1870) (see Table 4.1). The resolution of conflicting observations is adjudicated via application of the parsimony criterion, as discussed in Chapter 3. We will see in Chapters 6 and 7, under the discussion of character weighting, that additional logically consistent criteria have been advanced as further refinements of our attempts to unravel apparently conflicting observations at the level of homology.

There are many examples in the pre-cladistic literature of the a priori invocation of parallelism or convergence — that similar-appearing structures have developed from different precursor types. The arguments vary from example to example, but they are invariably not based on consistent evaluation of evidence.

As evidence of this ad hoc approach to interpreting structural variation consider the following example. Within the insect order Hemiptera, members of the family Miridae were grouped into subfamilies by Fieber (1860) primarily on the basis of morphological variation in the pretarsus — the claws and associated structures. In Fieber's system, and subsequent work of O. M. Reuter, the pretarsus was the single most important structural feature for subfamily recognition.

the nucleotide positions will match by chance alone if it is assumed that each of the four nucleotides occurs with equal frequency. Matching frequencies of greater than 25 percent could, then, be explained as the result of either homology or convergence.

The "statistical" viewpoint has been explicitly rejected by Mindell (1991), Brower and Schawaroch (1996), and others because it treats determination of sequence homology as a process of comparing the numbers of sites in common, an approach that is strictly phenetic. If the method of homology determination in DNA sequences is to correspond to that used for morphology, it is not the numbers of sites in common that is of interest, but to what degree we can find sites (or groups of them) that are unique across groupings of taxa. Because there are only four possibilities for change at any site, the problem of homology determination may be more difficult than in the case of gross morphology. Nevertheless, we are still observing changes at corresponding sites as they occur across taxa, not the sum of changes (differences) for a nucleotide string across taxa. In the end, correct matching should maximize the number of site correspondences, but the matches will be evaluated on a site-by-site basis, which would not be the case under the approach advocated by Patterson (1987).

At the level of the locus, as opposed to individual nucleotides, Fitch (1970) proposed, and others — including Patterson (1987) — have adopted, the term *paralogy* for use in reference to duplicated gene regions, those which exist as multiple similar, if not identical, copies. de Pinna (1991) noted that this usage corresponds to the usages of *serial homology, mass homology,* and *homonymy* in morphology; that is, for structures that occur in multiple copies in the same organisms. Paralogous genes, according to Fitch (1970), stand in contrast to *orthologous* genes in which "the history of the gene reflects the history of the species" (see further discussion in Sidebar 10, Gene Trees versus Species Trees). de Pinna concluded that there appears to be nothing special about homology in RNA-DNA sequence data as compared to morphology, only that new terms have been created to describe existing concepts.

Natural Groups

Groups that can be recognized on the basis of synapomorphy are referred to as "natural" or *monophyletic.* Hennig grasped the unequivocal importance of the concept of monophyly. Under his conception (see also Nelson, 1971; Farris, 1974), such a group has the following attributes:

• A *monophyletic* group contains the common ancestor and all of its descendants. Such groups are characterized by the possession of synapomorphies. One might conclude that if all groups recognized in a classification must be monophyletic, then there should be no need to distinguish among non-monophyletic groups. Hennig, nonetheless, recognized two other types of groups.

The works of Fieber and Reuter gained wide influence. The resp
worker W. E. China observed (China, 1933) that the structural in
the pretarsus had been exaggerated and that it was "far too plastic
fundamental group character." The argument, as couched by subse
ers, was that similarity of structure was the result of adaptation t
habits — convergence. The source of structural variation in the pre
Miridae is unknown. But, in the view of W. E. China, similarity of 1
necessarily imply common origin. To this day, however, pretarsal st
used to recognize systematic groupings within the 10,000 recognize
Miridae, not only because of the discrete variation they manifest but a
that variation is congruent with the distribution of other features in

Discussions of parallelism and convergence are no longer so (
as they once were. Whereas in the past the existence of these pher
often identified a priori, it is now understood through the results
analyses.

Homology Concepts as Applied to DNA-RNA Sequence Data

Before we leave the discussion of homology, the issue of nucleotid
data should be addressed. Whereas homology statements have tradi
volved morphological structures of at least moderate complexity (
morphological landscape, nucleotide sequence data do not readily sa
of these requirements. Nonetheless, analysis of sequence data depen
ries of site homology and nucleotide transformation.

DNA and RNA are composed of only four nucleotides, the purin
and guanine, and the pyrimidines cytosine and thymine (uracil in RN
fore, the question of whether two or more sites that occur in apparer
topographical positions are actually homologous — simply because t
cupied by the same nucleotide — may be less confidently judged tha
the wings of a bird are homologous with the forelimbs of a non-volant
To be certain, the equivalence of nucleotide (sequence) position in th
of different organisms is at times far from self-evident. Some parts
nome, such as protein-encoding regions, contain similar numbers of n
across a broad range of taxa. Other parts, such as ribosomal DNA,
variable numbers of nucleotides across a range of taxa, with some area
some areas duplicated, or anomalous segments inserted. This type of
complicates the task of comparing the nucleotide sequences among t
sible approaches to dealing with this issue are discussed in Chapter 5 (
heading of *alignment*.

The problem of determining homology in DNA and amino acid s
data has been erroneously described as a statistical problem by Patters(
and others. Under this characterization, if "identical" portions of the
from two different organisms are written out side by side, about 25 p(

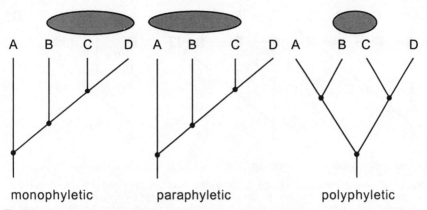

monophyletic   paraphyletic   polyphyletic

Fig. 4.4. Monophyletic, paraphyletic, and polyphyletic groups. In the sense of Hennig's definitions given in the text, ancestors are represented by the interior nodes of the cladogram.

- A *paraphyletic* group contains the common ancestor and some — but not all — of its descendants. Such groups are characterized by the possession of plesiomorphies.
- A *polyphyletic* group contains some of the descendants of a common ancestor but not the common ancestor itself. Such groups are characterized by the possession of convergent characters.

These definitions can be visualized in Fig. 4.4.

Hennig's definitions, although generally workable, were not consistently character based, as are definitions subsequently proposed by Farris (1974):

- A group is *monophyletic* if its group membership character appears uniquely derived and unreversed.
- A group is *paraphyletic* if its group membership character appears uniquely derived but reversed.
- A group is *polyphyletic* if its group membership character appears non-uniquely derived.

The character distributions associated with Farris' definitions are shown in Fig. 4.5.

The monophyly concept of Hennig is thus in accord with the theory of descent with modification, grouping together all members belonging to a lineage, and excluding none of them. As is discussed elsewhere in this book, when classifications contain paraphyletic and polyphyletic groups, the descriptive capacity of the characters used to define groups contained within those classifications is compromised. Thus, the recognition of monophyletic groups not only brings classifications into agreement with genealogy, if that is what we take them to mean, but serves the broader scientific purpose of recognizing groups for which there is maximum evidential support.

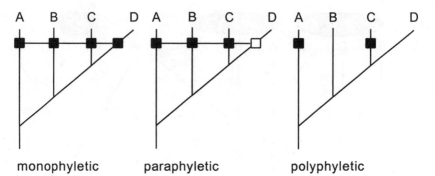

Fig. 4.5. Character distributions for groups under the definitions of Farris (1974). Black boxes indicate the distribution of the uniquely derived states for a character; open boxes indicate reversals. In the monophyletic group A + B + C + D, all taxa possess the group-defining feature in its uniquely derived and unreversed condition. That same group becomes paraphyletic when the condition of the feature is reversed in taxon D, and D is therefore not included as a member. A polyphyletic group is formed when the features uniting taxa A + C are not uniquely derived.

Sidebar 4
The Varied Conceptions of Monophyly

Monophyly has been defined by some authors as "a group all of whose members are descended from a single ancestor." So phrased the definition allows for virtually any grouping to be monophyletic. Willi Hennig and others used monophyly in a more restricted sense to mean "all descendants of a single (common) ancestor." The merits of both definitions have been argued on the basis of historical precedence. Monophyly in the sense used by Hennig is thought by most authors to be the same as that of Haeckel (1866), and because Haeckel first used the term, it is this usage that should have precedence.

The North American authors Simpson, Mayr, and Ashlock argued that monophyly had long been applied to groups that in Hennig's sense are paraphyletic, but that those groups nonetheless merit formal recognition. These authors invoked precedence on the basis of long-standing usage, but less for reasons of methodological consistency than for retaining groupings already existing in classifications. As part of one man's attempt to resolve this argument, Ashlock (1971) proposed the term *holophyly,* to replace Hennig's monophyly. Although not widely used, the term still has its adherents.

You might ask, "What is it that distinguishes monophyletic and paraphyletic groups in the system of definitions used by Ashlock"? The answer seems to be, nothing!

Determining Character Polarity

The homology concept, then, allows us to postulate *transformation series*—comprising two to several states (conditions) for any feature — that can have *polarity* and, if the number of states is more than two, *order*. It is polarity that we will discuss first; order will be discussed in Chapter 5.

Rooting by Outgroup Comparison

Polarity, or rooting, is determined by the choice of an *outgroup,* or *root*. This approach was implied in the works of Hennig, although not under this terminology. Using the traditional Hennigian approach, polarity was usually determined on a character-by-character basis during the data gathering phase, and individual character polarity determination was the subject of extended discussion in papers from the 1970s and early 1980s. Computer algorithms, such as the Wagner algorithm originally described by Farris, minimize the number of character state changes among taxa without regard to the polarity of the characters. The hierarchic structure, or network, created by the algorithm is then rooted by specifying one (or more) of the taxa as the outgroup (Fig. 4.6; see further discussion in Chapter 6). Under this formalization of cladistics, all of the literature describing methods to determine individual character polarity really addresses a nonproblem. It is only the choice of the outgroup that matters.

Outgroup comparison, as modified from the description of Nixon and Carpenter (1993), proceeds as follows:

1. Define or circumscribe ingroup taxa on the basis of presumed synapomorphies.
2. Select outgroup(s) on the basis of synapomorphies at a more inclusive level (higher level of generality), usually as based on a higher-level cladistic analysis.
3. Perform unrooted parsimony analysis.
4. Root cladogram between outgroup(s) and ingroup.
5. Read character polarities from cladogram.

In some cases, no known group may serve as an obvious candidate for the outgroup. In such cases the root for the ingroup should be determined as occurring along the internode that provides the most parsimonious tree length for the hypothetical ancestor and the ingroup (Nixon and Carpenter, 1993).

You might imagine that states for all characters are derived for the ingroup, or at least some of its members, and that in the outgroup all states are primitive. Although less common, the condition for a given character in the outgroup may also be derived. Such characters will add no information to the phylogenetic analysis unless they show variation within the ingroup.

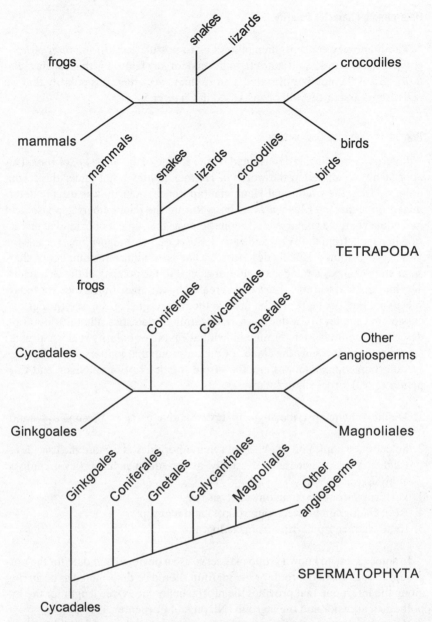

Fig. 4.6. Unrooted networks (above) for Tetrapoda and Spermatophyta, and their respective clado-grams (below). Polarities of characters used to form the networks are determined by choice of an outgroup. Frogs are chosen to root the tetrapod cladogram because, among other attributes, they lack the distinctive fetal membranes found in all other members of the group. Cycads are chosen to root the seed plant cladogram because, among other attributes, they lack the axillary branch-ing found in all other seed plants.

Sidebar 5
Discredited Methods of Determining Character Polarity

Several methods for determining character polarity have been proposed in addition to outgroup comparison and the ontogenetic criterion. These approaches have largely been abandoned by modern-day phylogeneticists in favor of logically based criteria related to minimization of homoplasy. These discredited methods include the following:

- *Common equals primitive.* This criterion suggests that more widespread characters are primitive relative to characters of more restricted distribution. It has no empirical justification.
- *More complex characters are derivative relative to less complex characters.* The subjectivity of this criterion would seem self-evident. The basis for complexity is seldom defined, let alone understood from a developmental or genetic point of view. Indeed, reduction and loss frequently seem to represent derived conditions.
- *Characters found in fossils are primitive relative to those found in living taxa.* The early rejection of the "paleontological method" in cladistics forced the abandonment of this approach, going hand in hand with the recognition of fossils simply as additional taxa subject to analysis and at most as indicating the minimum age for a taxon (see excellent discussions in Schoch, 1986).
- *Chorological progression.* In this view, the more advanced characters (and taxa) are to be found further from the geographic center of origin for a group. This "progression rule," which was advocated by Hennig and some other authors (e.g., Brundin, 1981), rests on the unwarranted assumption that "primitive" and "advanced" taxa can be recognized and that what might be true of one taxon in this regard should also be true of others (see Platnick, 1981).

Ontogenetic Data and Character Polarity

Outgroup comparison, as described above, has sometimes been referred to as the "indirect method" of determining character polarity. Hennig (1966:95) explicitly recognized the potential value of ontogenetic data in determining character polarity, and Nelson (1973, 1978) subsequently proposed that such data offer a "direct method" for determining character polarity. Nelson (1978) stated the case for ontogenetic data as follows:

> Given an ontogenetic character transformation, from a character observed to be more general to a character observed to be less general, the more general character is primitive and the less general advanced.

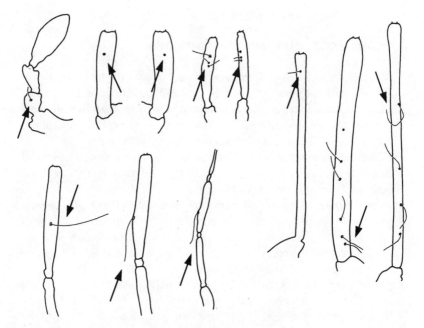

Fig. 4.7. Distribution of trichobothria (indicated by arrows) on antennal segment 2 in the nymphs and adults of some species of Pachynomidae and Reduviidae (from Wygodzinsky and Lodhi, 1989).

As an example, gill slits occur in the embryos of all vertebrates and in the adults of relatively primitive vertebrate groups (fishes) but not in the adults of more advanced vertebrates (tetrapods). According to Nelson's dictum, one may therefore assume that the more general condition (possession of gills) represents the primitive condition, and the modification of gills into other structures in adults represents the less general — derived — condition.

In order to better understand the nature of ontogenetic data, let us examine one of many possible available examples, that of the antennal trichobothria in the assassin bugs of the families Pachynomidae and Reduviidae (Insecta: Hemiptera: Heteroptera). Trichobothria are specialized setae that occur widely in the Heteroptera (Fig. 4.7). They are found on the antennae in only a few Gerridae and all Pachynomidae and Reduviidae, in all post-embryonic life stages. Virtually all Heteroptera have five nymphal stages, after which they become adults. It is known that some morphological details differ between the first instar and the later nymphal instars and that these same details may differ again in the adult, thus offering potential evidence of ontogenetic transformation.

The antennal trichobothria of Pachynomidae and Reduviidae were surveyed by Wygodzinsky and Lodhi (1989). The patterns of ontogenetic and between-group variation demonstrated in their work are summarized in Table 4.2. Al-

Table 4.2. Numbers of antennal trichobothria on antennal segment 2 in the Pachynomidae and Reduviidae

Taxon	Instar 1	Instars 2–5	Adult
Pachynomidae (*Aphelonotus*)	—	—	1
Reduviidae			
Cetherinae (*Eupheno*)	—	1	5
Harpactorinae			
Amphibolus	3	4	4
Arilus	—	11	12–15
Castolus	3	—	9
Heniartes	—	7	12
Notocyrtus	6	—	8
Phymatinae	1	1	1
Reduviinae (*Leogorus*)	—	1	8
Salyavatinae (*Salyavata*)	1	1	10
Triatominae (*Triatoma*)	—	1	10

Source: Data from Wygodzinsky and Lodhi (1989).

though the data are not complete for all taxa listed, we might draw the following conclusions:

1. the general condition is 1 trichobothrium, this situation occurring in all nymphal instars and adults of at least some taxa;
2. the general condition does not occur in all first instar nymphs, nor in all nymphs;
3. number of trichobothria increases during progression through the life stages in some taxa; and
4. trichobothrial numbers apparently never decrease during ontogeny.

Nelson (1978) stated that he knew of no examples that would falsify his restatement of Haeckel's "biogenetic law," which in its original formulation postulated that ontogeny recapitulates phylogeny. Nelson's proposal that more general characters are always primitive and less general characters are always advanced has been viewed in three different ways: true, partly true, and false.

Weston (1988) agreed with Nelson's conclusion that the direct method occupies the logically fundamental position in cladistic analysis because it provides information on synapomorphy without recourse to previous cladistic analyses. Weston interpreted some criticisms leveled at Nelson's approach as little more than redefining terms in order to obviate the argument. He reasoned that ontogenetic information provides the most basic synapomorphy hypotheses in cladistics but, in apparent contradiction to Nelson's stance, allowed that the information may not be infallible. This point had been made 6 years earlier by Rosen (1982:78), who noted that:

As a practical matter the problem of encountering incongruence among ontogenetic character transformations within a given taxon must be addressed. It matters little whether the cladistic disagreement posed by such incongruence is due to real reversals ... or to analytic failure; the decision as to which transformations specify the true hierarchy (of character states or organisms) must be decided, as in all cladistic disagreement, by the parsimony criterion.

Q. D. Wheeler (1990), using an empirical example from beetle ontogeny, reasoned that *character adjacencies* — the positions of character states relative to one another — can be directly observed in ontogeny but polarities cannot, and that patterns of character distributions can be observed, but their causal processes cannot. Wheeler viewed Nelson's rule as describing an indirect method for determining character polarity because it relies on a second taxon to estimate polarity. Furthermore, in Wheeler's view, successful polarization depends on the second taxon that is chosen for comparison. Wheeler concluded that Nelson's rule is a special case of parsimony and has the advantage of allowing for conclusions based on fewer comparisons than would be necessary using character data derived from only a single life stage. He also noted, as had Weston, that there are some situations in which ontogenetic data fail to polarize characters unequivocally, as is the case with simple outgroup comparison.

Nelson's assertion that there are no falsifiers of the restated biogenetic law was attacked by Kluge (1985) on the assumption that contradictory ontogenies are known. Kluge further criticized Nelson's view because he interpreted "more general" to mean commonality, with the suggestion that the argument could be reduced to "common equals primitive." However, as pointed out by Weston (1988), the more general character is possessed by all taxa that possess the less general character as well as some that do not; thus, Kluge's equation of common with primitive is incorrect (see also Sidebar 5, Discredited Methods of Determining Character Polarity).

Let us examine the use of ontogenetic data for character-polarity determination in light of the example in Table 4.2 for reduviid antennal trichobothria.

Can polarity be determined directly? Our example suggests that ontogenetic data cannot determine polarity on the basis of a single taxon. If we had only two taxa, one with a single antennal trichobothrium in all life stages (Phymatinae) and one with multiple trichobothria in all life stages (*Amphibolus*), the question of what condition was general — and therefore primitive — could not be answered. Lacking information on the Pachynomidae, the sister group of the Reduviidae, one still might not come to an unequivocal conclusion concerning the general condition in the Reduviidae, depending on the sample of taxa available. Thus, Wheeler's conclusion, that ontogenetic data are a special case outgroup comparison because polarity determination requires more than one taxon would seem to be correct.

Are there any falsifiers? The answer to this question would seem to have two parts. The patterns of trichobothrial distribution, as far as they are known, do not contradict the general conclusion that a single trichobothrium is the general condition. That is, there are no known instances in which there are multiple trichobothria in the nymphal stages and a single trichobothrium in the adult. However, the general condition does not occur universally in nymphs, because as seen in Table 4.2 first instar nymphs of *Amphibolus, Castolus,* and *Notocyrtus* have multiple trichobothria. What does appear to be regular is that if the numbers of trichobothria change during ontogeny, they always increase — an observation concordant with the generally observed terminal addition of traits during ontogeny. Variation in ontogenetic data for the Reduviidae is similar to that described by Wenzel (1993) for the ontogeny of nest-building behaviors in the paper-wasp subfamily Polistinae (see Chapter 10).

In sum, ontogenetic data appear to represent an important source of characters for phylogenetic analysis. They may contain ambiguities, as do other forms of data. Ontogenetic data, where available, should be coded like any other. Whether or not such data offer a direct method of polarity determination, they offer a clear-cut indication of state ordering, a subject discussed in Chapter 5.

Choice and Formulation of Outgroups

The efficacy of outgroup comparison is related to the attributes of the taxa chosen to serve as outgroups. Outgroup choice for most morphological studies has been predicated on relationships established in prior higher-level studies.

The outgroup need not be limited to a single taxon, and indeed, analyses are probably best conducted using multiple outgroups. Multiple taxa serving as outgroups do not have to form a monophyletic group (Nixon and Carpenter, 1993). Some analyses have used a single representative species as the outgroup; others have used a composite taxon that combines characteristics found across a range of species. The use of actual species makes justification of character coding more straightforward than is often the case with a composite hypothetical outgroup. In the latter case, character codings — of necessity — will be "deduced" in the outgroup rather than observed.

A distant or otherwise poorly chosen outgroup for rooting analyses of DNA sequence data will possess few sites in common with the ingroup taxa and therefore serve its intended function poorly. W. C. Wheeler (1990) pointed out that an essentially random collection of nucleotides will usually root a phylogeny on the longest branch or internode; he concluded that such an outcome is unreliable or even meaningless. A valid rule of thumb might be that if an outgroup could not root a tree effectively on the basis of morphology, there should be no reason to believe that the same taxon would be any more effective in the rooting of a tree based on sequence data.

Ancestors, Sister Groups, and Age of Origin

The taxa, including fossils, in Hennig's phylogenetic schemes (and all subsequent cladistic work) were always placed as terminals at the ends of branches, whereas many preceding phylogenetic approaches embedded taxa, particularly fossils, along the branches of the tree (see Fig. 4.8). You might ask: What are *ancestors* in the sense of Hennig's definition of monophyletic? Can we actually find them in nature? Are they represented by fossils? And, if they cannot be found in nature, does this not discredit the entire system?

Cladistic methods treat ancestors as hypothetical entities (sometimes called "hypothetical taxonomic units," or HTUs), which have an inferred "ground plan" of attributes. Ancestors, like all unobservable entities, are inferences based on the results of systematics; they are not something that can be recognized directly. Whether some fossil actually represents an ancestor seems to be a question that resides in the realm of the unknowable. Thus, fossils are terminal taxa, like all other taxa in phylogenetic analysis (see, e.g., Hennig, 1969). This approach stands in stark contrast to the "paleontological approach," where fossils are de facto ancestors of other recognized taxa, either those in geologic strata of lesser age, or of Recent forms.

The logic of cladistics demands that all groups be monophyletic. Thus, if one taxon arises from another, the "ancestral" taxon is, by definition, paraphyletic. From the point of view of character distributions, ancestor–descendant relationships are untestable in a cladistic framework because, as pointed out by Engelmann and Wiley (1977), corroboration of such hypotheses would be based on plesiomorphy. Ancestors would be of necessity taxa recognized by the absence of the distinctive attributes of their descendants. Therefore, virtually any grouping of putative ancestors and descendants would be possible.

Nelson applied the term *cladogram* only to diagrams depicting recency of common ancestry, whereas he restricted the term *tree* to diagrams containing ancestor–descendant relationships (Fig. 4.8). In the present work the term *tree* is sometimes used interchangeably with *cladogram,* in conformity with much of the literature. This usage is not meant to suggest, however, that such "trees" are thought to transmit information on ancestor–descendant relationships.

Hennig referred to two (or more) groups arising from a common branch point in a cladogram as *sister groups* (Fig. 4.9). Because sister groups share the same hypothetical common ancestor, they are, by extension, of the same age.

Phylogenetic hypotheses in the form of cladograms, then, offer us information on the relative age of origin of taxa. When fossil taxa are among those being analyzed, there may be reason to assume a minimum age of origin on the basis of the fossil evidence. Cladistic methods do not, however, provide information on absolute age of origin, nor do they allow for the identification of taxa as ancestors. Neither do cladograms represent *direct* depictions of the history of speciation. Their limitations in this regard are both practical and methodological.

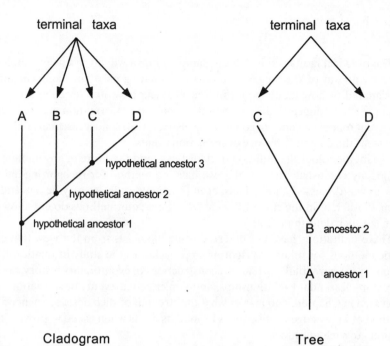

Cladogram   **Tree**

Fig. 4.8. Comparison of cladograms and trees in the sense of Nelson, showing the nature of ancestors in the two types of diagrams. Ancestors in cladograms are hypothetical, represented by the nodes that connect groupings of terminal taxa. The characters attributed to those nodes are optimized (as described in Chapter 6) from the characters of the terminals. These deduced attributes form the "ground plan" for the hypothetical ancestor, or hypothetical taxonomic unit (HTU). Ancestors in the tree are taxa represented by actual specimens, usually fossil.

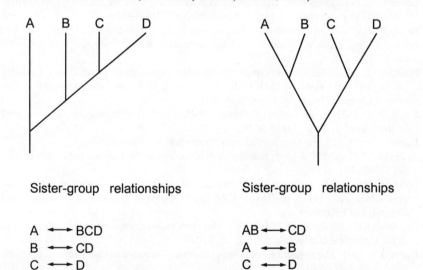

Fig. 4.9. Sister-group relationships identified for four taxa on two cladograms with different topologies. Sister groups are deemed to be of equal age because they share the most recent common ancestor.

The practical limitation is that our sample of taxa will always be incomplete. We have no way of knowing how many species have gone extinct and will never be sampled or how many from the Recent fauna have not yet been sampled. Therefore, although a cladogram based on character information may correctly represent relative recency of common ancestry, it will in most cases certainly not represent the complete history of speciation events.

The methodological limitation of cladistics — and therefore of any method of phylogeny reconstruction — is that we have no method for recognizing ancestors, as was discussed above. Thus, even though one species may be descended from another, as in the case of Recent from fossil, our methods do not allow us to make such determinations.

These limitations have not deterred some biologists from the view that cladistic methods (or other less rigorous approaches to the study of relationships among taxa) can actually produce reconstructions of evolutionary history, reveal ancestor–descendant relationships, and depict actual events of speciation. As discussed in Chapter 3, no matter what the strength of such desires, science as a method of knowledge acquisition is beyond its limits when asked to provide the answers to such questions.

Literature Cited

Ashlock, P. D. 1971. Monophyly and associated terms. *Syst. Zool.* 20:63–69.

Boyden, A. 1947. Homology and analogy. A critical review of the meaning and implication of these concepts in biology. *Amer. Midl. Nat.* 37:648–669.

Brady, R. H. 1985. On the independence of systematics. Cladistics 1:113–126.

Brower, A. V. Z., and V. Schawaroch. 1996. Three steps of homology assessment. *Cladistics* 12:265–272.

Brundin, L. Z. 1981. Croizat's panbiogeography versus phylogenetic biogeography. pp. 94–138. *In:* Nelson, G., and D. E. Rosen (eds.), *Vicariance Biogeography: A Critique.* Columbia University Press, New York.

China, W. E. 1933. A new family of Hemiptera–Heteroptera with notes on the phylogeny of the suborder. *Ann. Mag. Nat. Hist.* ser. 10, 12:180–196.

Darwin, C. 1859. *On the Origin of Species.* John Murray, London. 490 pp.

de Pinna, M. C. C. 1991. Concepts and tests of homology in the cladistic paradigm. *Cladistics* 7:367–394.

Eldredge, N., and S. J. Gould. 1972. Punctuated equilibria: an alternative to phyletic gradualism. pp. 82–115. In Schopf, T. M. (ed.), *Models in Paleontology.* Freeman and Cooper, San Francisco.

Engelmann, G. F., and E. O. Wiley. 1977. The place of ancestor-descendant relationships in phylogeny reconstruction. *Syst. Zool.* 26:1–11.

Farris, J. S. 1971. The hypothesis of nonspecificity and taxonomic congruence. *Ann. Rev. Ecol. Syst.* 2:277–302.

Farris, J. S. 1974. Formal definitions of paraphyly and polyphyly. *Syst. Zool.* 23:548–554.

Fieber, F. X. 1860. *Die europaischen Hemiptera.* Halbflugler (Rhychota Heteroptera). Carl Gerold's Sohn, Wien. 444 pp.

Fitch, W. M. 1970. Distinguishing homologous from analogous proteins. *Syst. Zool.* 19:99–113.

Geoffroy St. Hilaire, E. 1818. *Philosophie anatomique.* Paris.

Haeckel, E. 1866. *Generelle Morphologie der Organismen.* G. Reimer, Berlin.

Hennig, W. 1965. Phylogenetic Systematics. *Ann. Rev. Entomol.* 10:97–116.

Hennig, W. 1966. *Phylogenetic Systematics.* University of Illinois Press, Urbana. 263 pp.

Hennig, W. 1969. *Die Stammesgeschichte der Insekten.* Waldemar Kramer, Frankfurt am Mein. 436 pp.

Kluge, A. G. 1985. Ontogeny and phylogenetic systematics. *Cladistics* 1:13–27.

Lankester, E. R. 1870. On the use of the term homology in modern zoology, and the distinction between homogenetic and homoplastic agreements. *Ann. Mag. Nat. Hist.* (4)6:34–43.

Lipscomb, D. L. 1992. Parsimony, homology, and the analysis of multistate characters. *Cladistics* 8:45–65.

Mayr, E. 1982. *The Growth of Biological Thought: Diversity, Evolution, and Inheritance.* Belknap Press of Harvard University Press, Cambridge, Massachusetts, London.

Mindell, D. P. 1991. Aligning DNA sequences: homology and phylogenetic weighting. pp. 73–89. *In:* Miyamoto, M. J., and J. Cracraft (eds.), *Phylogenetic Analysis of DNA Sequences.* Oxford University Press, New York.

Nelson, G. 1971. Paraphyly and polyphyly: redefinitions. *Syst. Zool.* 20:471–472.

Nelson, G. 1973. The higher-level phylogeny of vertebrates. *Syst. Zool.* 22:87–91.

Nelson, G. 1978. Ontogeny, phylogeny, paleontology, and the biogenetic law. *Syst. Zool.* 27:324–345.

Nixon, K. C., and J. M. Carpenter. 1993. On outgroups. *Cladistics* 9:413–426.

Owen, R. 1843. *Lecture on the Comparative Anatomy and Physiology of the Invertebrate Animals.* Longman, Brown, Green, and Longman, London.

Patterson, C. 1982. Morphological characters and homology. pp. 21–74. *In:* Joysey, K. A., and A. E. Friday (eds.), *Problems in Phylogenetic Reconstruction.* Academic Press, London.

Patterson, C. 1987. Introduction, pp. 1–22. *In:* Patterson, C. (ed.), *Molecules and Morphology in Evolution: Conflict or Compromise?* Cambridge University Press, Cambridge.

Platnick, N. I. 1981. Discussion of: Croizat's panbiogeography versus phylogenetic biogeography, by L. Z. Brundin. pp. 144–150. *In:* Nelson, G., and D. E. Rosen (eds.), *Vicariance Biogeography: A Critique.* Columbia University Press, New York.

Remane, A. 1952. *Die Grundlagen des Naturlichen Systems der Vergleichenden Anatomie und der Phylogenetik.* Geest und Portig K. G., Leipzig.

Rieppel, O. C. 1988. *Fundamentals of Comparative Biology.* Birkhauser, Basel. 202 pp.

Rosen, D. E. 1982. Do current theories of evolution satisfy the basic requirements of explanation? *Syst. Zool.* 31:76–85.

Schoch, R. M. 1986. *Phylogeny Reconstruction in Paleontology.* Van Nostrand Reinhold, New York. 353 pp.

Wenzel, J. W. 1993. Application of the biogenetic law to behavioral ontogeny: a test using nest architecture in paper wasps. *J. Evol. Biol.* 6:229–247.

Weston, P. H. 1988. Indirect and direct methods in systematics. pp. 25–56. *In:* Humphries, C. J. (ed.), *Ontogeny and Systematics.* British Museum (Natural History), London. 236 pp.

Wheeler, Q. D. 1990. Ontogeny and character phylogeny. *Cladistics* 6:225–268.

Wheeler, W. C. 1990. Nucleic acid sequence phylogeny and random outgroups. *Cladistics* 6:363–367.

Wygodzinsky, P., and S. Lodhi. 1989. Atlas of antennal trichobothria in the Pachynomidae and Reduviidae (Heteroptera). *J. New York Entomol. Soc.* 97:371–393.

Suggested Readings

de Pinna, M. C. C. 1991. Concepts and tests of homology in the cladistic paradigm. *Cladistics* 7:367–394. [A useful review of homology concepts]

Maddison, W. P., M. J. Donoghue, and D. R. Maddison. 1984. Outgroup analysis and parsimony. *Syst. Zool.* 33:83–103. [A helpful discussion of outgroups and optimization]

Nixon, K. C., and J. M. Carpenter. 1993. On outgroups. *Cladistics* 9:413–426. [An up-to-date review of outgroup comparison]

Patterson, C. 1982. Morphological characters and homology. pp. 21–74. *In:* Joysey, K. A., and A. E. Friday (eds.), *Problems in Phylogenetic Reconstruction.* Academic Press, London. [Within the modern literature, a classic paper on homology concepts]

Rieppel, O. C. 1988. *Fundamentals of Comparative Biology.* Birkhauser, Basel. 202 pp. [An excellent discussion of the philosophical underpinnings of homology concepts]

Weston, P. H. 1994. Methods for rooting cladistic trees. pp. 125–155. *In:* Scotland, R. W., D. J. Siebert, and D. M. Williams (eds.), *Models in Phylogeny Reconstruction.* Clarendon Press, Oxford. [A useful discussion of outgroup comparison]

5

Character Analysis and Selection of Taxa

The selection of characters and taxa lays the groundwork for all subsequent steps in phylogenetic analysis. The process of observing, coding, and rechecking character information has been referred to as "character analysis." Although central to phylogenetic analysis, there are divergent opinions about what the process represents and how it should proceed. In this chapter we will examine character selection, character coding, DNA sequence alignment, and issues relating to the selection of taxa.

What Is a Character?

Characters comprise any heritable attributes that possess group-defining variation, particularly those that show congruence with other such features. Traditionally, morphology at the macroscopic level has formed the basis for most recognized taxonomic characters. More recently, DNA and amino-acid sequences have become "standard" character sources for many groups, augmenting classical morphology. Behavior and products of behavior also enjoy a place as legitimate sources of character data (Michener, 1953; Wenzel, 1992). Characters may range from simple to complex.

Character information does not present itself directly. It is not self-revealing. Character data is, rather, a synthesis of observation. Thus, there is the potential for divergence of opinion among investigators about how to recognize the "characters" possessed by a given group of organisms and concerning the possible approaches for representing that information in a form suitable for analysis.

Character Analysis

If we had only two taxa under study, the question of their relationship to one another would be trivial — no matter the number and distribution of characters between them — because, of necessity, they would be each others' closest relatives. Multiple characters for three or more taxa allow for the testing of grouping schemes among those taxa. Characters that show no variation among three or more taxa, or are unique to terminal taxa, are useless for resolving relationships among the constituent taxa because they imply no grouping schemes.

The application of the positivist philosophy of science to systematics would suggest that once the characters are observed, all subsequent decision making should follow directly. The philosophy of science more generally associated with cladistics indicates a reciprocal relationship between the character data and cladograms, treating both characters and cladograms as theories subject to test.

It is through the process of "coding" that information on characters is made available for use in phylogenetic analysis. However, because the characters themselves do not tell us how they should be recognized and coded, we must establish some criteria by which to convert observations into codes. If, as stated by Pimentel and Riggins (1987), a character includes all homologous expressions of a feature found in the ingroup and outgroup, then the codes used to represent the various observed conditions of the feature would be called *states*. Brower and Schawaroch (1996) observed that characters and their states become formalized as different things through the construction of data matrices, whereby characters are represented by columns of cells and states are represented by the values in the individual cells. Because any given state may serve as a group-defining feature, the terms *character* and *state* are often used interchangeably, depending on the hierarchic level at which they apply. Platnick (1978:366), for example, was of the opinion that "different character states are merely homologies that are synapomorphic at different levels, and there is thus a hierarchy of increasingly restricted characters rather than a division of characters into alternate states."

Coding Criteria

Under the assumption that each expression of a feature is unique to a single group, one might envision a system in which all characters are coded as present in one group and absent from all other groups. So coded, characters would represent observations, or conjectures, and the only test would be congruence with other characters. Thus conceived, character theories are presumed by some authors to be minimally theory laden. A character theory could then be rejected when the character appears in more than one group on the cladogram or, as stated by Lipscomb (1992:55), when there is "non-congruence due to non-homology in two or more different taxa." This is *scattering* (Fig. 5.1) in the parlance of Mickevich and Lipscomb (1991).

For those features that only manifest two conditions in the groups being studied — as many features do — the question of how they should be coded is usually

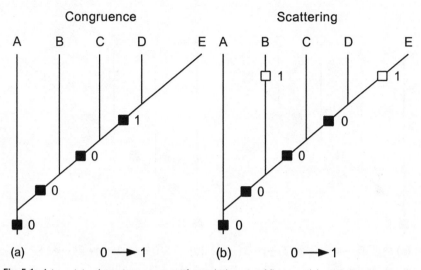

Fig. 5.1. A two-state character, congruent for a cladogram of five taxa (a), and showing scattering, with the derived condition occurring independently in more than one terminal taxon (b).

not at issue. The conditions are either "this" or "that"; for example "wings absent or wings present" in insects or "seeds absent or seeds present" in plants. The question of how characters should be coded often arises when there are three or more conditions for a feature, such as "tarsi with 2, 3, 4, or 5 segments" or "corolla with 3, 4, 5, 6, or numerous petals."

Mickevich (1982) has argued strenuously against treating all character data in a two-state (presence–absence) format, urging instead the use of multistate codings when three or more conditions of a homologous feature exist. Under this view, multistate codings are said to represent more meaningful and bolder hypotheses of character transformation. The degree to which multistate-character theories disagree with the cladogram they define is termed *hierarchical discordance* (Fig. 5.2; Mickevich and Lipscomb, 1991), which is the "non-congruence due to non-homologous similarity used to hypothesize order of states in multistate characters" (Lipscomb, 1992:56).

The implementation of multistate codings requires an approach for postulating transformation from one state to another. If the characters (homologs) are recognized on the basis of similarity of position and structure, then it is the "connection by intermediates" that allows hypothesizing state-to-state transformation. At least four techniques have been proposed for coding such transformational information.

- *Non-additive* or *Fitch transformation* (Fitch, 1971) codes multistate characters with all conditions of a homologous feature being treated as related, and then analyzes them in such a way that any state may transform to any other state

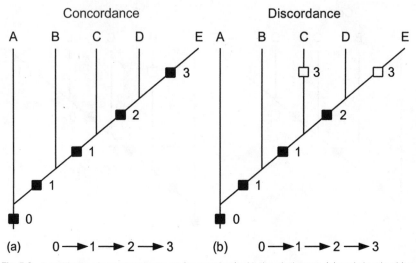

Fig. 5.2. A multistate character congruent (concordant) with the cladogram (a) and showing hierarchical discordance (b), character-state 3 not fitting the cladogram perfectly.

with equal "cost" (Fig. 5.3; see also Optimization, Chapter 6). This approach is most commonly used in parsimony analyses of DNA sequence data, data that would otherwise have to be analyzed under some specified model of sequence evolution. Use of non-additive coding achieves maximal agreement between cladograms and the characters on which they are based. However, in the case of morphological characters, this agreement is achieved by ignoring observed information on transformation (see example in Fig. 6.4). All other coding approaches imply state-to-state transformational order.

- *Morphocline analysis* (Maslin, 1952), as here defined, orders states solely on the basis of similarity, without regard for congruence, and may therefore produce transformation series in which homoplasious similarity is confused with homology (Lipscomb, 1992). Under this precladistic approach, rules other than congruence would be invoked to adjudicate character conflicts.

- *Transformation series analysis* (TSA) (Mickevich, 1982) orders character states on the basis of similarity. The hypothesized transformation is then tested by whether or not the character is concordant (congruent) with the cladogram it defines, that is, whether the character-state transformations and the cladogram are in agreement. If not, the character is recoded, and the data are reanalyzed. This process is repeated iteratively until agreement among all multistate characters and the cladogram is maximized. This coding method might be envisioned as an extension of Hennig's idea of "checking, correcting, and rechecking." TSA was criticized by Lipscomb (1992:61) for discarding information on similarity among states in favor of codings that define a more resolved cladogram. Because no algorithm exists for checking

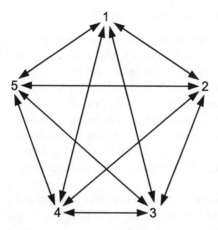

nonadditive, unordered, Fitch transformation

additive, ordered, Farris transformation

Fig. 5.3. Diagrammatic representation of unordered (non-additive; Fitch) transformation (above), which allows any state-to-state change and is frequently used in the analysis of DNA sequence data. Ordered (additive; Farris) state-to-state character transformation (below) as ordinarily used in morphocline analysis, transformation series analysis, and homology analysis. Under the non-additive approach, no state-to-state theories of transformation are implied, and any state-to-state transformation is possible without the imposition of additional steps. Under the additive approach, transformations from one state to another are postulated on the basis of observed similarities in the states, and additional transformational steps are required if the state-to-state order does not conform to the cladogram.

global concordance for all multistate characters on a given cladogram it may be impossible to tell whether an optimal result has been achieved under this method.

- *Homology analysis* uses both similarity and congruence to evaluate character codings. Lipscomb (1992) argued that the flaws of the three above-mentioned approaches to multistate coding can be avoided by employing this method. When using homology analysis, information on observed similarity is not discarded simply because the character has an imperfect fit to the cladogram. Rather, Lipscomb recommended checking all similarity-based transformation theories for congruence with other data and accepting those showing maximal congruence. She, however, rejected recoding similarity-based transformations simply to achieve congruence.

Example and Discussion of Character Coding

If the forelimbs of mammals were judged to be homologous with the wings of birds, few would doubt the theory, and it might even seem justified without recourse to other character information. On the other hand, the theory of limb loss in snakes might be thought stronger via recourse to congruence with other characters because most snakes possess no structural remnants of the forelimbs and only a few snakes show hind limb remnants. All of these conditions might also be judged homologous with the pectoral fins of fishes. In *presence–absence format* the fore-limb characters could be coded as four characters, where "0" represents absent and "1" represents present, the unique condition:

1. 0) anterior appendages absent, 1) anterior appendages present as fins
2. 0) anterior appendages not in the form of limbs, 1) anterior appendages in the form of limbs
3. 0) anterior appendages not present as wings, 1) anterior appendages present as wings
4. 0) anterior appendages present, 1) anterior appendages absent (lost)

The same information could be coded in *multistate format* as a single character, where the numbering sequence indicates the adjacencies of the states:

0) anterior appendages absent, 1) anterior appendages present as fins, 2) anterior appendages present as limbs, 3) anterior appendages present as wings, 4) anterior appendages absent (lost)

Graphically the multistate coding can be presented as follows:

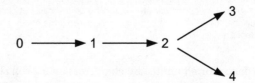

Wilkinson (1995), in an examination of character coding approaches, coined the terms "reductionist coding" and "composite coding." In the above example, the presence–absence coding could be termed *reductionist coding,* treating as many conditions as possible as unrelated. This type of coding was referred to by Pimentel and Riggins (1987) as the "stepwise decisions approach," which they said offered the utility of pencil and paper solutions, but which could introduce errors into the analysis. The approach was defended by Pleijel (1995), who believed it to be simpler and more straightforward than multistate coding because it removes the necessity for making decisions about state order, allowing that information to emerge as a result of cladistic analysis. In spite of the benefits envisioned by Pleijel, hypotheses of transformational similarity are not retained when a feature with three or more states is coded in presence–absence format.

The multistate coding shown above represents the *composite coding* of Wilkinson. This approach may combine conditions that some observers would say are not homologous and that therefore should be coded via a reductionist approach. In the above example, leg loss in some amphibians, some lizards, and snakes might be grouped — incorrectly — as a single character "anterior appendages absent." Yet, to its advantage, multistate coding treats "all homologous expressions of a feature" as parts a single character, thereby preserving information on similarity and transformation.

Because the test of synapomorphy is congruence among characters, we might then wish to code all character data to maximize congruence. Whether this is best achieved via two-state or multistate codings is not always clear.

Nonetheless, presence–absence coding seems to have the following limitations:

- The codes for some states may not represent the result of observations. This phenomenon will probably occur most frequently for certain "absence" states in characters that would have otherwise been coded as multistate.
- Absence connotes a nongroup forming condition of a feature. Therefore, presence–absence coding implies knowledge of polarity a priori. For many attributes, such knowledge may appear virtually self-evident, as for example the relative recency of origin of tetrapod limbs versus fins in fishes, or the presence versus absence (notwithstanding secondary loss) of wings in insects. Yet, for many homologous features, polarity is far from obvious, especially in those cases where detailed cladistic analysis ultimately suggests character reversal.
- Presence–absence codings may obscure information on transformational homology.

Multistate codings, on the other hand, simply hypothesize possible transformations across multiple states (i.e., ordering) on the basis of similarity. This approach to coding does not necessarily imply polarity. The optimal rooting, and therefore polarity of a character, will be determined by addition of the outgroup.

The assumptions of the coding methods described above can be described as follows:

1. Presence–absence coding does not even assume that alternative states are manifestations of the same character concept.
2. Nonadditive coding assumes a common character concept, but does not assume a particular transformation (Fig. 5.3).
3. The remaining methods (morphocline analysis; transformation series analysis; homology analysis), all of which represent additive codings, make both assumptions (Fig. 5.3).

On the basis of the above discussion, it can be argued that features with three or more conditions should be coded as multistate. The value of such an approach is that (1) all state codings will be based on observation, and (2) information on observed transformational similarity will be preserved in the coding.

Data Matrices and Coding Methods

Character data, when being prepared for processing by computer algorithm, are normally assembled in the form of a matrix (Fig. 5.4). By convention the taxon names are listed as the rows on the left and the characters are listed as the columns across the top. Some data sets could be coded adequately and accurately based only on the discussion up to this point. Other data will contain complexities that require knowledge of additional techniques, however.

Linear versus Branching Characters

Multiple states of a character can be coded in a linear format, up to a total of 10, for most computer programs. Such coding may take on a branching form if

```
Character       0     5    10    15    20    25    30    35    40    45    50    55    60    65
Hypochilus      00001 11000 00000 00000 00000 00000 00000 00000 00000 00100 00001 -0000 00000 00
Ectatosticta    00001 11000 00000 00000 00000 00000 00000 00010 00000 00100 00000 -0000 00000 00
Gradungula      00010 00111 11020 00100 00000 00000 00000 00001 00001 00000 0000? -?000 00000 00
Pianoa          00010 00111 11020 00100 00000 00000 00000 00001 00001 00000 0000? -?000 00000 00
Hickmania       00010 00111 10110 00100 00000 00000 00020 00001 00001 00000 0000? -1000 00000 00
Austrochilus    00010 00111 10110 01100 00000 00000 00020 00000 00001 00000 0000? -1000 00000 00
Thaida          00010 00111 10110 01100 00000 00000 00020 00000 00000 00000 00001 -1000 00000 00
Filistata       00110 00101 10001 12000 00000 00000 10011 10001 00001 10000 00010 -1000 00010 11
Kukulcania      00110 00101 10001 12000 00000 00000 10011 10001 00010 10000 00010 -1000 00010 11
Scytodes        11110 00111 10001 12000 00000 00000 -0141 10101 0000- -0000 00000 -0010 01010 00
Sicarius        11110 00111 10001 12000 00000 00000 -0141 11101 1000- -0000 00000 -0000 00010 00
Drymusa         11110 00111 10001 12000 00000 00000 -0131 10101 0000- -0000 0000- -0010 01010 00
Loxosceles      11110 00111 10001 12000 00000 00000 -0141 11101 1000- -0000 0000- -0100 01010 00
Diguetia        11110 00111 10001 12000 00000 10001 -0121 10101 1000- -0000 00000 -200- 11010 10
Segestrioides   11110 00111 10001 12000 00000 00000 -0121 10101 1000- -0000 00000 -2001 1101? 00
Plectreurys     01110 00111 10001 12000 00001 10000 -0121 10111 100-- -0000 00000 -210- 1111? 00
Kibramoa        01110 00111 10001 12000 00001 10000 -0121 10111 100-- -0000 00000 -2100 1111? 00
Pholcus         01110 00111 10001 12000 00000 00000 -0141 10101 110-- -0000 00000 -1001 1111? 00
Caraimatta      11110 00111 10001 12010 00000 00000 -0020 10111 100-- -0000 0000- -1000 00110 00
Nops            01110 00111 10001 12010 00000 00000 -0041 10101 1000- -0000 00000 -1000 01010 00
Ochyrocera      11110 00111 10001 12000 00000 00000 -0130 10101 1001- -0000 00000 -1000 01010 00
Segestria       11010 00111 10201 12010 00000 00000 -0040 10011 0000- -0000 00000 -1000 0001? 00
Dysdera         11010 00111 10201 12010 00000 00000 -0040 10011 1000- -0000 00000 -1000 00010 00
Mallecolobus    11010 00111 10201 12010 00000 00000 -0040 11011 0000- -0000 00000 -?000 00010 00
Dysderina       11010 00111 10201 12010 00000 00000 -0040 11111 1000- -0000 00000 -1000 00010 00
Appaleptoneta   11010 00111 10001 12000 00010 00000 -1140 10001 1001- -0000 00000 00000 00000 00
Usofila         11010 00111 10001 12000 00010 00000 -1040 10101 1001- -0000 00000 00000 00000 ?0
Archaea         01010 00111 10001 12001 11110 00000 -0030 00010 1010- -0001 10000 01000 00001 10
Mecysmauchenius 01010 00111 10001 12001 111?0 00000 -0130 00110 1010- -0001 10000 ??000 00001 00
Tricellina      01010 00111 10001 12001 11010 00000 -0140 00110 1010- -0000 00000 00000 00000 10
Huttonia        01010 00111 10001 12001 10010 00000 -0140 00010 1010- -0000 00000 11000 00001 10
Otiothops       01010 00111 10001 12001 100?0 01000 -0140 00010 1010- -0000 00000 ??000 00001 10
Waitkera        00010 00-11 10001 12000 00010 00010 00130 00010 10-0- -0000 01002 21000 00000 ?0
Tetragnatha     01010 00111 10001 12000 00010 00110 -0130 00010 10-0- -0100 00103 01000 0000? ?0
Crassanapis     01010 00111 10001 12000 00010 00110 -0130 00100 10-0- -0000 00103 01000 00000 10
Oecobius        00010 00111 10001 12000 00010 00000 00100 00010 10000 11000 00010 01000 00001 01
Stegodyphus     00010 00111 10001 12000 00010 00000 00100 00010 00-00 11000 00012 01000 00001 10
Deinopis        00010 00111 10001 12000 00010 00010 00100 00010 00-01 01000 01002 21000 00001 01
Dictyna         00010 00111 10001 12000 00010 00000 00140 00010 10101 01010 0000? 21000 00000 ?0
Callobius       00010 00111 10001 12000 00010 00000 00120 00010 10101 11010 00002 11000 00001 10
Araneus         01010 00111 10001 12000 00010 00000 -0130 00010 00-01 -1100 01000 01000 00000 10
Mimetus         01010 00111 10001 12001 10010 00000 -0130 00010 0010- -1100 00000 01000 00000 10
Pararchaea      01010 00111 10001 12001 111?0 00000 -0130 00100 1010- -1100 0000? ??000 00000 10
```

Fig. 5.4. A typical data matrix, with the first taxon (*Hypochilus*) representing the outgroup. Missing information is represented by a dash (from Platnick et al., 1991: Table 2; courtesy of The American Museum of Natural History).

the condition found in the outgroup is placed somewhere in the middle of the sequence of states, rather than at either end, as shown below.

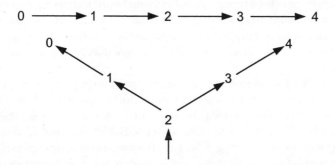

This approach was labeled *internal rooting* by O'Grady and Deets (1987), who observed that although it minimizes the number of variables (columns in the data matrix) required to code a simple branching multistate character, it has the disadvantage of showing the outgroup condition as something other than the conventional zero. In the example, 2 becomes the outgroup condition in the internally rooted character, whereas it would be zero in the linear character.

Not all computer phylogenetics programs currently in use directly support complex branching in character–state trees. Thus, if you wish to code character data which take on the form of more complex branching character–state trees, you must do so with techniques such as *additive binary coding,* or *nonredundant linear coding* (see Farris et al., 1970; O'Grady and Deets, 1987; Pimentel and Riggins, 1987; O'Grady, Deets, and Benz, 1989). These methods will allow for any character–state tree to be described accurately. However, such recoded characters will occupy more than one column in the matrix, complicating comprehension of the relationship of the various states of the character in synapomorphy lists and other diagnostic output from the programs. Both types of coding maintain the state-to-state order assigned at the time of coding. Characters need not be coded in non-additive formats, even if one wishes to process the data under that approach to optimization, because all computer phylogenetics programs allow, as an option, the processing of additively coded characters on a non-additive basis. The converse is not true, however; therefore, it is always preferable to code characters in a format that includes hypothesized state-to-state transformations based on observation. This caveat does not apply to DNA sequence data.

Additive binary coding. This technnique requires variables (characters; columns) equal in number to the number of terminals plus the number of internal nodes for the tree, if the character has a distinctive condition for all of the taxa being coded. If the character does not possess a distinctive condition for every taxon being coded, the number of variables necessary to code the information will be reduced. The matrix values are determined by assigning a "1" (ingroup condition) code to each terminal contained in the most inclusive grouping on the tree. The process is then repeated for the second most inclusive grouping on the tree, and so on, until all inclusive groupings have been coded. As with any coding technique, autapomorphies are unique to the terminals that possess them, and therefore a "1" (derived) code will be assigned as a unique value for each terminal.

An example of additive binary coding is shown in Fig. 5.5. Six hypothetical taxa have the multistate character transformation shown in Fig. 5.5a. The character–state distributions among the taxa are shown in Fig. 5.5c. The matrix shown in Fig. 5.5b contains the five binary variables required to recover the branching pattern shown in Fig. 5.5c. The groupings formed by each variable are shown on the cladogram and above each variable in Fig. 5.5b. No autapomorphic states of this character exist in any of the taxa, and therefore none are coded in the matrix.

Nonredundant linear coding. This technique requires fewer variables than additive binary coding because multistate coding is used to code the states for the longest ordinal subset (longest sequence of states) on the character–state tree. Additional binary variables are then used to code the branches that remain. For

Sidebar 6
Character Independence

The literature contains many statements arguing that character independence is a prerequisite to conducting phylogenetic analysis. As such, independence is simply a requirement of method in science. Cladistics is looking for groups supported by evidence in the form of characters, a point made by Hennig when he observed that the more characters defining a group, the greater our confidence in the monophyly of the group. If characters lacked independence, confidence in groups defined by multiple characters would not vary from that for groups defined by single characters.

Structures existing on different parts of the body would appear, de facto, to represent independent characters. But, for characters that are structurally associated, there may be no direct evidence for independence; rather independence could be assumed if the structurally allied characters serve to define groupings of taxa at different levels in the cladogram. Arguments for independence have been advanced for DNA data derived from different parts of the genome and known to have different functions, and for DNA and gross morphological data.

this reason, the method has also been referred to as "ordinal-additive binary coding" and "mixed coding" by Pimentel and Riggins (1987). Returning to the example character–state tree shown in Fig. 5.5a, we see one possible approach to nonredundant linear coding for this hypothesized transformation in Fig. 5.5d. Variable 1, now in multistate form, codes for the nodes on the cladogram using four conditions. The branch comprising taxa B + C is coded in variable 2. No autapomorphic states of this character exist in any of the taxa, and therefore none are coded in the matrix. A total of 2 variables code the same data that required 5 variables using additive binary coding.

Autapomorphies

One might assume that a matrix of character data would always include as much information as possible. Consider, however, autapomorphies, characters unique to terminal taxa. We might wonder whether or not they should be included in data matrices used for phylogenetic analyses. On the one hand, it has been argued that autapomorphies should not be included because they inflate the value of "fit statistics," such as the consistency index (discussed in Chapter 6), and therefore make comparisons of such statistics less meaningful. On the other hand, if the taxa being analyzed are not species, but rather collections of species, the autapomorphies add information by documenting the monophyly of the terminals even though they offer no information on grouping of those terminals with other taxa.

We might conclude that if the goal of a study is to make comparisons of the amount of homoplasy among matrices, then removal of autapomorphic characters would be a desirable approach. If, however, the primary goal of assembling data matrices is part of a continuing effort to document the distinctive attributes of the taxa under study as well as the monophyly of supraspecific terminal taxa, then autapomorphies should be included in the matrix. Computer phylogenetics programs make it possible to compute trees and fit statistics using only informative characters. Thus, all attributes should be included in a data matrix. The uninformative characters can easily be excluded for the purposes of computation without physically modifying the data matrix.

Missing Data

There are two types of absences for which transformational information is truly not available. These have been referred to as *missing* (unavailable) and *inapplicable* data.

Missing data, in the sense used here, are those characters that cannot be coded for a given taxon because the information is not available at the moment. The

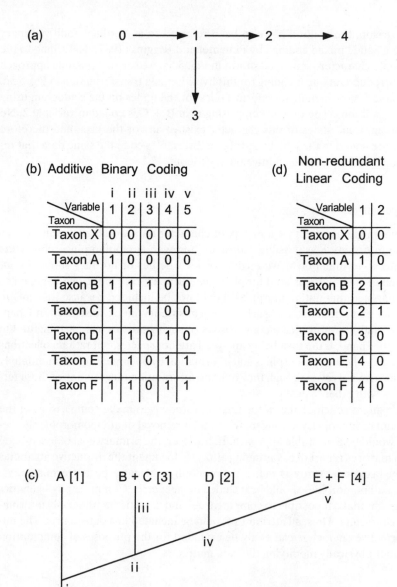

Fig. 5.5. A hypothetical branching character–state tree (a) coded in additive binary format (b), and in nonredundant linear format (d). The relationships of the taxa derived from the two coding approaches are shown in the cladogram (c). The states from the character state tree (a) are shown in brackets on the cladogram (c). The inclusive groupings for which the additive binary variables code are indicated in parts *b* and *c* by small Roman numerals. See text for further explanation.

data might become available in the future with the acquisition of more specimens, or specimens in a proper state of preservation. *Inapplicable data* are those
that can never be coded for a given taxon.

An example may help to illustrate the difference between these data types.
Most insects have wings, and the wings themselves possess many distinctive attributes useful for grouping taxa in which they occur. But, secondarily wingless
insects cannot be coded for attributes of the wings themselves, as for example,
patterns of venation. Therefore, attributes of wings are inapplicable in secondarily wingless taxa. On the other hand, some species of insects exist in both winged
and wingless morphs. If a matrix were coded using the wingless morph because
only that form was available at the time, the relevant states should be coded as
though the data were unavailable, in lieu of acquiring winged specimens.

In light of the above discussion, we might conclude that it is desirable to distinguish between missing and inapplicable data in preparing a data matrix. Such
a distinction will facilitate future character analysis and additions of data to matrices. Computer phylogenetics programs do not distinguish between the two
types of absences during the computational process, however.

There may be many cells coded as missing in a matrix of data sets combined
for "simultaneous analysis," as, for example, with morphological and DNA sequence data collected by different investigators for the same higher taxon, but
that do not represent identical terminal taxa. Because it is known that missing
observations sometimes cause negative perturbations in the analysis, what approaches might we take to minimize the numbers of empty cells in the matrix?
Two available options are to *fuse terminals* or to use *exemplar terminals* (Nixon
and Carpenter, 1996). If it is acceptable to combine sequences from one terminal taxon with morphology from another because each partial data set is a fair
representation of variation for the combined taxa as a whole, terminal fusion is
an option. The nature of such fusions should be made clear under discussion of
data coding and analysis, to facilitate future testing of results, and to allow incorporation of new data that eliminates the requirement for fusing terminals. In
using the exemplar terminal approach, we might choose to eliminate some terminals represented by incomplete data, on the assumption that other included
terminals with complete data will fairly represent the higher taxon to which all
of the terminals in question belong.

Polymorphisms

Some authors have coded as missing, characters that are polymorphic in terminal taxa, the so-called X-coding of Doyle and Donoghue (1986). Nixon and
Davis (1991) criticized that approach because it does not take into account variation in the terminals during phylogenetic analysis; consequently, the resulting
cladograms are often not computed correctly. Therefore, polymorphic terminal

taxa should be separated into all possible combinations of character states that are known to occur within such lineages. This approach will allow for the discovery of parsimonious solutions in which polymorphic lineages may be found to be polyphyletic.

Intrinsic versus Extrinsic Data

A distinction between intrinsic and extrinsic data has been made by a number of authors. Intrinsic characters are those obviously subject to the mechanisms of inheritance, such as gross morphology and DNA sequences. Extrinsic characters are those not apparently subject to the rules of inheritance, as for example biogeographical and host associations. The former type of character data has traditionally been used in the construction of cladograms. The latter type has usually been examined to determine the degree to which it is congruent with intrinsic character data (see further discussion in Chapters 10 and 11).

Kluge and Wolf (1993) argued for the inclusion of "extrinsic" as well as "intrinsic" data in phylogeny reconstruction. They suggested that such an approach involves a "consilience of inductions." What Kluge and Wolf failed to acknowledge, in their advocacy of using extrinsic characters, is that there is no method for determining homology or transformation for extrinsic data, independent of the results of intrinsic-character–based phylogenetic results themselves. Therefore, there is no justification for treating such observations as part of a parsimony analysis. Kluge and Wolf quoted Mickevich (1982) in supporting their view, but they perverted the meaning of her statements to include character data beyond the type with which she was dealing.

Take, for example, host–parasite relationships, a widely studied phenomenon in the co-evolution literature, and a subject discussed in detail by Hennig. The sequence of host changes might be looked upon as a problem in co-speciation, whereby parasite speciation parallels host speciation. That assumption was made by Brooks (1981) in his proposal for a "solution to Hennig's parasitological method." However, the assumption would seem to be unwarranted because monophyletic groups of parasites are often observed to occupy distantly related hosts, with the clear implication that co-speciation was not involved; for example, in restricted monophyletic groups of phytophagous insects occurring on conifers and angiosperms, or legumes and composites, in the case of some plant-feeding insects.

Although host relationships, geographic occurrences, and ecological associations may all represent part of the history for a group of organisms, there are strong arguments for the view that the methodologies for assessing agreement among these data are different than those used to analyze data subject directly to the "laws" of inheritance, contrary to the views of Brooks (1981) and Kluge and Wolf (1993). Kluge (1997) remained undeterred in his view, asserting that host–parasite (co-evolutionary) and biogeographic relationships will serve as more critical tests of phylogenetic theories than synapomorphies themselves, a notion whose justification he did not explain further. Methods for evaluating extrinsic data are discussed further in Section III, Application of Cladistic Results.

Alignment of DNA Sequence Data

Determining homology of sites for comparable DNA–RNA sequences across a range of taxa is referred to as *alignment.* Alignment of protein-coding regions is relatively easy because of "the presence of structured reading frames with predictable features of more frequent change at third base positions within codons, and recognizable start and stop codons" (Mindell, 1991:75). Furthermore, the congruence criterion can be used to test theories of site homology, because if amino acid translations of sequences correspond to one another across taxa, this offers a measure of corroboration of the underlying nucleotide alignment. Straightforward alignments, such as those found in protein-coding genes, are often done "visually." Indeed, there is the view — although certainly not universally held — that if alignment is not obvious, all further attempts at analysis of such data will be futile.

Non-protein-coding regions of the genome present distinct challenges to alignment because they vary in length from taxon to taxon, a condition brought about by insertions and deletions (indels). Thus, with only four possible nucleic acid occurrences at any given site, variation in the number of sites often makes determination of site homology across taxa less than clear-cut. Nonetheless, such regions, as for example 12S, 16S, and 18S ribosomal RNA and DNA, show useful variation at higher levels of relationships and therefore have been extensively used in phylogenetic analysis of a broad range of animal taxa. Corroboration of alignment via comparison with amino acid translations is not possible. Visual alignment is untenable for any data set with a large number of taxa and substantial sequence variation because there is no way to evaluate anywhere near the number of possible alignments without the use of a computer.

The crucial nature of determining positional homology among nucleotides, especially its determination in non-protein-coding genes, would seem to mandate that alignment procedures be treated as an integral part of the phylogenetic analysis of such data. Nonetheless, Hillis et al. (1996), in their volume dedicated to methods and techniques for molecular systematics admitted that alignment was difficult and poorly understood but nonetheless relegated the subject to a chapter dealing with sequencing and cloning under the subheading "Interpretation and Troubleshooting."

Proposed alignment methods are of several types. Some imply correspondence to prior conclusions about relationships and can be discarded as not treating sequence data as an independent source of evidence. The remaining methods can be divided into *similarity methods* and *parsimony methods.*

Alignment based on similarity has the same drawbacks, discussed in Chapter 1, found in grouping nonmolecular data by overall similarity. Nonetheless, analyses based on overall similarity are still applied by many investigators working with molecular data, and it is therefore not overly surprising to encounter the use of similarity-based alignment procedures, such as CLUSTAL (Higgins and Sharp, 1988).

The remaining approaches to alignment attempt to maximize the number of site correspondences through the application of the parsimony criterion. There are many ways that this problem might be solved. Confidence in having found the solution(s) optimal for the criterion used requires that the method be computer implemented. The first alignment programs dealt with two taxa at a time, using the so-called pairwise approach. The first attempt at parsimony-based multiple sequence alignment was that of Wheeler and Gladstein (1992) with the program MALIGN. Such an approach requires substantial computing power, but evaluates alignments on a global basis for all taxa under consideration, rather than by using sequential pairwise comparisons.

Alignment methods rely on the insertion of gaps to accommodate for the differences in numbers of nucleotide positions among the sequences being aligned (Fig. 5.6). Because gaps do not represent observations, Wheeler (1996) proposed a parsimony-based technique that he termed "optimization alignment" as one possible approach to obviating that drawback. Under this method, insertion–deletion events are treated not as states, but rather as transformations linking ancestral and descendant nucleotide sequences (see Chapter 6, for further discussion of optimization).

It should be noted that alignment algorithms often are designed to conform to some model of molecular evolution — such as the increased weighting of transversions relative to transitions — or can be tailored to do so. The effects of invoking one or more such models should be examined empirically with regard to maximizing character congruence and parsimonious interpretation of the data, as was done by Wheeler (1995), rather than being applied as a priori assumptions.

Other Types of Character Information

There is an extensive literature on types of character information other than those discussed so far. Many of the techniques for collecting and analyzing such data were conceived at a time before cladistic theory became the obvious method of choice in systematics. Some approaches have lost their appeal because of the methodological problems they present during analysis or because the data types have been largely supplanted by nucleotide sequences. Three of the less frequently encountered data types deserve brief discussion.

Anoplodactylus	––––––––TCCGC–CCCA–CCCGTGG–GGCCGGAGGC––––
Limulus	––––––––––C–C–GCC–––T–T–A–C–GAGGTGGGGC––––
Centruoides	––––––––––CGG–GGC–––T–C–T–T–GGCTCCGGGC––––
Mastigoproctus	––––––––––CGT–GCC–––GCGAG–ATCGGCACTCGA––––
Peucetia	––––––––––C–T–CCC–––G–GAG–A–CGGGACGGGC––––
Nephila	––––––––––CCG–TCT–––T–TGTCA–CTGACGGGGC––––
Balanus	––––––––––C–C–GTC–––C–T–G–T–GGGGCGGC–C––––
Scutigera	GTCTTCCCCTTACCCTTTCGGGGGTGGGAGTTGGCGGC
Spirobolus	––––––––––CGG–TGG–––G–ATTAG–ATCTCCGGGC––––
Libellula	––––––––––AGG–CTC–––T–TCG–C–GGAGGCTCCC––––
Mantis	––––––––––CTC–TCC–––T–CAA–C–GGGGGAGCCC––––
Papilio	––––––AGCGT–CGG–––TCGTTCGATCGGCCTCTC––––

Anoplodactylus	ATGCAGATCTTCGTGAAGACCTTGACCGGAAAGACCATCACTCTGGAAGT
Limulus	ATGCAGATCTTCGTGAAGACCTTGACCGGAAAGACCATCACTCTGGAAGT
Centruoides	ATGCAGATCTTCGTGAAGACCTTGACCGGAAAGACCATCACTCTGGAAGT
Mastigoproctus	ATGCAGATCTTCGTGAAGACCTTGACCGGAAAGACCATCACTTTGGAAGT
Peucetia	ATGCAGATCTTCGTGAAGACCTTGACCGGAAAGACCATCACTTTAGAAGT
Nephila	ATGCAGATCTTCGTGAAGACCTTGACCGGAAAGACCATCACATTAGAAGT
Balanus	ATGCAGATCTTCGTGAAGACCTTGACCGGAAAGACCATCACTCTTGAAGT
Scutigera	ATGCAGATCTTCGTGAAGACCTTGACCGGAAAGACCATCACCTTGGAAGT
Spirobolus	ATGCAGATCTTCGTGAAGACCTTGACCGGAAAGACCATCACCCTAGAAGT
Libellula	ATGCAGATCTTCGTGAAGACCTTGACCGGAAAGACCATCACTTTGGAGGT
Mantis	ATGCAGATCTTCGTGAAGACCTTGACCGGAAAGACCATCACTTTGGAAGT
Papilio	ATGCAGATCTTCGTGAAGACCTTGACCGGAAAGACCATCACCCTGGAAGT

Fig. 5.6. Alignment of sequence data. Two sets of sequences from the same group of 12 taxa. Above, aligned 18 ribosomal DNA sequences, showing the insertion of gaps represented by dashes to accommodate the variation in numbers of nucleotides across taxa. Below, a fragment of sequences from the nuclear protein-coding ubiquitin gene, showing the constant number of nucleotides and perfect site correspondence within the fragment (from Wheeler, Cartwright, and Hayashi, 1993; courtesy of W. C. Wheeler).

Frequency Data

The most commonly encountered frequency data come from allozymes. Such data are commonly used in the study of populations. They have also been used to some extent in a phylogenetic context. Allozyme data have three properties that impinge on the way they might be analyzed: (1) different taxa (or populations) may be polymorphic for a single electrophoretic product; (2) there is often little information in common between taxa, with nearly all information unique to taxa (or populations); and (3) "identical" electromorphs frequently represent heterogeneous groups of alleles at the level of DNA sequences (Barbadilla, King, and Lewontin, 1996). Allozyme data have generally been treated through the use of distances (see below) instead of as discrete characters. Mickevich and Mitter (1981) discussed three possible discrete-character approaches to analyzing allozyme data: the *independent allele model* in which each allele becomes a separate character; the *shared-allele model,* where each locus becomes a single character whose states are combinations of alleles; and a *systematic approach* in which separate enzymes are assumed to be characters, the states of which are defined to be the allelic combinations. It is the last approach that would seem to be most relevant if such data are to be used in cladistic analyses.

Distance Data

Data gathered directly in a distance format are immunological data, DNA hybridization (annealing) data, and some others (see discussion of distances in Chapter 1). Of all data types gathered directly as distances, DNA–DNA hybridization data, especially as promoted in the laboratory of Charles Sibley, have probably attracted the most attention.

The allure of molecular-level distance data is rooted in two ideas. First, data derived from the genome are thought by some to speak more directly to the issue of evolutionary propinquity and divergence than do data derived from study of the phenotype. Second, if the rate of divergence were constant and could be quantified, we would then have a measure of evolutionary rate, a so-called molecular clock.

On the negative side, at least three arguments have been made against the use of distance data. First, data collected as distances contain no information on homology, in spite of assertions to the contrary. Gould (1985) claimed that DNA–DNA hybridization techniques as pioneered by Sibley had solved the problem of distinguishing homologies from analogies. Contrary to Gould's assertion, what they do in fact is imply some measure of "difference" between all possible pairings of taxa, but they contain no direct indication of what that difference might be.

Second, the intentional conversion of discrete character data into distances deliberately discards evidence pertaining to the nature of similarity — in the form of morphological structure or nucleotide sequences — and consequently on kinship. Therefore, conversions of discrete character data into distances should be avoided (Farris, 1981:21).

Third, there are methodological issues that bring into question our ability to analyze most data collected directly as distances. As noted above, the greatest attraction of distance data is their presumed ability to depict degree of divergence in a clocklike manner. In order to analyze data under the rate-constancy assumption, the data and the method of analysis must be metric, that is, satisfy the triangle inequality (see Sidebar 8 for a general discussion of "metrics"). The data must also fit the more stringent requirement of being ultrametric. Farris (1981) compiled considerable evidence contradicting the metricity of immunological data and made it clear that there was no evidence for an appropriate model with which to analyze DNA hybridization data. He further noted that the measures often used in the analysis of distance data (such as Nei's and Rogers' distances) were not metric in their properties, and therefore any results derived from their use were meaningless when the data were analyzed under the assumption that they were metric.

Measurement Data

Some authors, particularly those advocating phenetic and statistically based techniques, have argued for the use of measurement data, often under the heading of "morphometrics." The seeming merit of such an approach would be the "relatively objective" nature of the data themselves. Yet, measurements are not necessarily so easily interpreted, and it would be naive to assume that they do not potentially suffer from errors of accuracy and observation. Possibly the most important limitation of measurement data for phylogenetic analysis, however, is that measurements by themselves do not imply homology. Even if we admit to their utility, there exists little agreement on how to code continuous variables in a discrete-character format.

Measurements, in the form of ratios, can be useful for describing the general attributes of form, such as whether a given structure is longer than wide, or the relative lengths of structures within or among taxa. Morphometric techniques are also potentially capable of offering descriptions of complex shapes. Nonetheless, in spite of the nearly total quantification of cladistics, the contribution of morphometrics to the analysis of phylogenetic relationships is still decidedly marginal.

Selection of Taxa and Specimens

Up to this point in the chapter, characters have been the primary focus. Yet, the selection of taxa and specimens for systematic studies has implications of equal importance. Yeates (1995) recognized what he called "exemplar" and "ground-plan" (intuitive) methods for representing taxa in coded character data.

Under the *exemplar method,* actual taxa serve as the terminals in the analysis, and the data are coded directly from them. If the terminals are species, coding should be straightforward. If the terminals are higher taxa, it is important that they possess only a single state for all characters being coded. If two or more states are present in the terminal, then two or more exemplars are necessary, as was discussed above under polymorphisms. The greater the variation in a terminal taxon, the greater the number of exemplars necessary to allow for correct optimization of the characters for the hypothetical common ancestor of the group.

If terminal taxa are not species, we are faced with the question of how to diagnose them. The *ground-plan* (intuitive) method assigns character-state data to a terminal taxon based on examination of the literature or a range of specimens. The ground plan might be based on optimization of characters from a prior analysis. It must be "deduced" if no such analysis exists. Character assignments for a deduced ground plan will be unproblematic if there is no variation for the character in the terminals. If two or more states exist for a given character in a terminal taxon, that character must be coded as polymorphic in order to faithfully portray all character change among the taxa. Otherwise, "deduced groundplan" assignments can easily produce errors of optimization — that is, assignment of character-state sets for hypothetical common ancestors — for the analysis as a whole (Nixon and Davis, 1991).

The above considerations apply primarily to higher-level analyses, where the terminal taxa are not species or where species were chosen to represent supraspecific taxa. When conducting species-level studies, such as revisions, the most important consideration will be to acquire the broadest possible sample of taxa and specimens representing them. This will allow for corroboration or refutation of past observations as well as for the discovery of new information.

The search for specimens will most likely be guided by using sources described in Chapter 1. In addition to corroborating previously used characters and discovering new ones, examination of the broadest possible sample of specimens will also allow for the discovery of previously unrecognized, and therefore undescribed, taxa as well as the creation of new synonymy where necessary. The credibility of the results will be a reflection of the thoroughness of the search for relevant specimens.

Literature Cited

Barbadilla, A., L. M. King, and R. C. Lewontin. 1996. What does electrophoretic variation tell us about protein variation? *Mol. Biol. Evol.* 13:427–432.

Brooks, D. R. 1981. Hennig's parasitological method: a proposed solution. *Syst. Zool.* 30:229–249.

Brower, A. V. Z., and V. Schawaroch. 1996. Three steps of homology assessment. *Cladistics* 12:265–272.

Doyle, J. A., and M. J. Donoghue. 1986. Seed plant phylogeny and the origin of the angiosperms: an experimental approach. *The Botanical Review* 52:321–431.

Farris, J. S. 1981. Distance data in phylogenetic analysis. pp. 3–23. *In: Advances in Cladistics. Proceedings of the First Meeting of the Willi Hennig Society.* New York Botanical Garden, Bronx, New York.

Farris, J. S., A. G. Kluge, and M. J. Eckhardt. 1970. A numerical approach to phylogenetic systematics. *Syst. Zool.* 19:172–191.

Fitch, W. M. 1971. Toward defining the course of evolution: minimum change for a specific tree topology. *Syst. Zool.* 20:406–416.

Gould, S. J. 1985. A clock of evolution: We finally have a method for sorting out homologies from "subtle as subtle can be" analogies. *Natural History* 94:12–25.

Higgins, D. G., and P. M. Sharp. 1988. CLUSTAL: a package for performing multiple sequence alignment on a microcomputer. *Gene* 73:237–244.

Hillis, D. M., C. Moritz, and B. K. Mable (eds.). 1996. *Molecular Systematics,* second edition. Sinauer Associates, Sunderland, Massachusetts.

Kluge, A. G. 1997. Testability and the refutation and corroboration of cladistic hypotheses. *Cladistics* 13:81–96.

Kluge, A. G., and A. J. Wolf. 1993. Cladistics: What's in a word? *Cladistics* 9:183–199.

Lipscomb, D. L. 1992. Parsimony, homology, and the analysis of multistate characters. *Cladistics* 8:45–65.

Maslin, T. P. 1952. Morphological criteria of phyletic relationships. *Syst. Zool.* 1:49–70.

Michener, C. D. 1953. Life history studies in insect systematics. *Syst. Zool.* 2:112–118.

Mickevich, M. F. 1982. Transformation series analysis. *Syst. Zool.* 31:461–468.

Mickevich, M. F., and D. Lipscomb. 1991. Parsimony and the choice between different transformations of the same character set. *Cladistics* 7:111–139.

Mickevich, M. F., and C. Mitter. 1981. Treating polymorphic characters in systematics: a phylogenetic treatment of electrophoretic data. pp. 45–58. *In:* Funk, V. A., and D. R. Brooks (eds.), *Advances in Cladistics. Proceedings of the First Meeting of the Willi Hennig Society.* New York Botanical Garden, Bronx, New York.

Mindell, D. P. 1991. Aligning DNA sequences: homology and phylogenetic weighting. pp. 73–89. *In:* Miyamoto, M. J., and J. Cracraft (eds.), *Phylogenetic Analysis of DNA Sequences.* Oxford University Press, New York.

Nixon, K. C., and J. M. Carpenter. 1996. On simultaneous analysis. *Cladistics* 12:221–241.

Nixon, K. C., and J. I. Davis. 1991. Polymorphic taxa, missing values and cladistic analysis. *Cladistics* 7:233–241.

O'Grady, R. T., and G. B. Deets. 1987. Coding multistate characters, with special reference to the use of parasites as characters of their hosts. *Syst. Zool.* 36:268–279.

O'Grady, R. T., G. B. Deets, and G. W. Benz. 1989. Additional observations on nonredundant linear coding of multistate characters. *Syst. Zool.* 38:54–57.

Pimentel, R. A., and R. Riggins. 1987. The nature of cladistic data. *Cladistics* 3:201–209.

Platnick, N. I. 1978. Classifications, historical narratives, and hypotheses. *Syst. Zool.* 27:365–369.

Platnick, N. I., J. A. Coddington, R. R. Forster, and C. E. Griswold. 1991. Spinneret morphology and the phylogeny of haplogyne spiders (Araneae, Araneomorphae). *Amer. Mus. Novitates* 3016:73 pp.

Pleijel, F. 1995. On character coding for phylogeny reconstruction. *Cladistics* 11:309–315.

Wenzel, J. W. 1992. Behavioral homology and phylogeny. *Annu. Rev. Ecol. Syst.* 23:361–381.

Wheeler, W. C. 1995. Sequence alignment, parameter sensitivity, and the phylogenetic analysis of molecular data. *Syst. Biol.* 44:321–331.

Wheeler, W. C. 1996. Optimization alignment: the end of multiple sequence alignment in phylogenetics? *Cladistics* 12:1–9.

Wheeler, W. C., and D. G. Gladstein. 1992. *MALIGN: A Multiple Sequence Alignment Program.* Program and documentation. Vers. 2.0. The American Museum of Natural History, New York.

Wheeler, W. C., P. Cartwright, and C. Y. Hayashi. 1993. Arthropod phylogeny: a combined approach. *Cladistics* 9:1–39.

Wilkinson, M. 1995. A comparison of methods of character construction. *Cladistics* 11:297–308.

Yeates, D. K. 1995. Groundplans and exemplars: paths to the tree of life. *Cladistics* 11:343–357.

Suggested Readings

Farris, J. S. 1985. Distance data revisited. *Cladistics* 1:67–85. [An extended discussion of the properties of distance data and the difficulties involved in analyzing them]

Lipscomb, D. L. 1990. Two methods for calculating cladogram characters: transformation series analysis and the iterative FIG/FOG method. *Syst. Zool.* 39:277–288. [Approaches to computing "cladogram characters"]

Lipscomb, D. L. 1992. Parsimony, homology and the analysis of multistate characters. *Cladistics* 8:45–65. [A concise explanation of morphological character analysis]

Mickevich, M. F., and D. Lipscomb. 1991. Parsimony and the choice between different transformations of the same character set. *Cladistics* 7:111–139. [A useful discussion of character analysis and the relationship between characters and cladograms]

6

Quantitative Cladistic Methods

Although systematists have long associated characters with taxa, the relationship between character data and "phylogeny" has not always been obvious. The writings of Willi Hennig clarified this relationship. The formalization of concepts, via the parsimony criterion, has allowed for computer implementation of methods of phylogenetic inference and the realistic solution of heretofore intractable problems. It is this capability that has taken the study of taxonomic relationships from an almost purely qualitative enterprise to one that is dominated by the use of computer software. In this chapter we will examine "quantitative cladistics" in detail, including the issues of fit, parsimony algorithms, and character weighting. We will also further discuss the proposed use of maximum-likelihood techniques as an alternative approach to parsimony.

Background

Phylogenetic analysis, as proposed and practiced by Hennig, involved a restricted number of taxa and a limited amount of homoplasy in any given data set. In other words, the characters used often straightforwardly identified a unique scheme of relationships. For this reason, Hennig's method was at times misapprehended as implying clique analysis, an approach in which groups are formed only on the basis of characters that are all perfectly congruent (see Sidebar 7). The practical drawback of the strict Hennigian approach is that simple calculation often does not discover the multiple phylogenetic solutions that exist for a data set containing incongruent characters. The power of the computer offers the only possibility for solving problems in phylogenetic analysis other than those represented by small, highly consistent data sets.

Before considering the details of quantitative cladistic methods, it will be helpful to understand how character consistency can be measured on phylogenetic trees.

Fit Statistics

The term "fit" has been widely used in the phylogenetic literature to indicate the degree to which data conform to a cladogram. The most commonly used measure of fit applied to discrete character data is the *consistency index*, or *ci*, originally proposed by Kluge and Farris (1969). This is a simple measure, but one of considerable value. It is computed as follows (Fig. 6.1):

$$ci = \frac{m}{s}$$

where s is the observed number of character-state changes (steps) and m is the minimum number of such changes. Values for *ci* range from 1.00 for a perfect fit to near 0.00 for the worst possible fit.

The number of homoplasious steps in a character is then:

$$s - m = h \text{ (homoplasy)}$$

The consistency index, as a measure, may be applied to a single character or to the suite of characters on a cladogram; the latter measure (CI) was termed the *ensemble consistency index* by Farris (1989) and is the sum of the minimum number of possible changes for all characters divided by the sum of the observed number of changes for all characters:

$$CI = \frac{\Sigma m}{\Sigma s}$$

Because the consistency index does not have a minimum value of zero, Farris (1988; see Farris, 1989) introduced the *rescaled consistency index* (rci), which has a value of zero when a character has as much homoplasy as possible, a desirable feature when used in conjunction with some character-weighting functions. Computing the rci requires determining the maximum possible number of steps, g, for a character on a cladogram. This can be accomplished by the method shown in Fig. 6.1, whereby the number of possible changes for a two-state character is equivalent to the number of terminal taxa in which the character can change. The rci is then computed as follows:

$$rci = \frac{g - s}{g - m} \cdot \frac{m}{s}$$

Mickevich and Lipscomb (1991) argued that *concordance* between character-state transformations and the hierarchy of the tree is a better measure than the consistency index of how well multistate characters fit a cladogram. However,

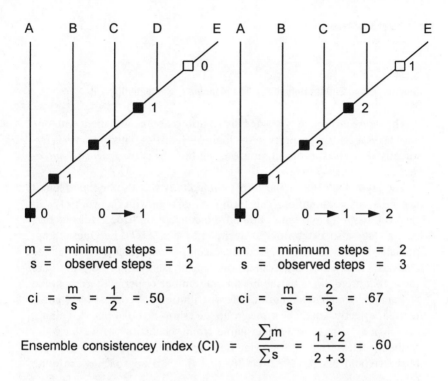

$$m = \text{minimum steps} = 1 \qquad\qquad m = \text{minimum steps} = 2$$
$$s = \text{observed steps} = 2 \qquad\qquad s = \text{observed steps} = 3$$

$$ci = \frac{m}{s} = \frac{1}{2} = .50 \qquad\qquad ci = \frac{m}{s} = \frac{2}{3} = .67$$

$$\text{Ensemble consistencey index (CI)} = \frac{\sum m}{\sum s} = \frac{1+2}{2+3} = .60$$

Computing g, maximum possible number of steps:

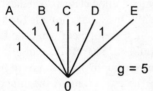 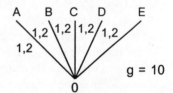

Rescaled consistency index (rci):

Example 1
$$rci = \frac{g-s}{g-m} \cdot \frac{m}{s} = \frac{5-2}{5-1} \cdot \frac{1}{2} = .375$$

Example 2
$$rci = \frac{g-s}{g-m} \cdot \frac{m}{s} = \frac{10-3}{10-2} \cdot \frac{2}{3} = .58$$

Fig. 6.1. Computing consistency indices. For the two-state character $0 \rightarrow 1$, the minimum possible number of steps, m, on a cladogram is one, allowing change from state 0 to state 1. In the example, the character shows a reversal to state 0 in taxon D, therefore requiring a total of 2 observed steps, s, on the cladogram. The consistency index is then computed as $ci = m/s = 1/2 = .50$. For the Three-state character $0 \rightarrow 1 \rightarrow 2$, the minimum possible number of steps, m, is 2, allowing changes from state 0 to state 1, and from state 1 to state 2. In the example, the character shows homoplasy in state 1, therefore requiring a total of 3 steps on the cladogram. The consistency index is computed as $ci = m/s = 2/3 = .67$.

The maximum possible homoplasy, g, for a character can be determined as shown, by plotting the character on a "bush." The rescaled consistency index (rci) can then be computed as shown.

Sidebar 7
Similar Terms, Similar Definitions: The Meaning of Compatibility

The degree to which character distributions are in agreement with one another is a central property in the logic of cladistics. This property is frequently described in the literature using the terms *congruence, concordance,* and *consilience.*

The etymologically similar term *compatibility* will also be found; but it has a distinct connotation. This term is associated with the work of Walter Le Quesne (1969) who outlined the concept of "true cladistic characters" (see also Estabrook, Strauch, and Fiala, 1977). Le Quesne's approach, which is usually referred to as "compatibility analysis" or "clique analysis," is based on the assumption that the best supported groups are those that are defined by the maximum number of perfectly compatible characters, or maximal cliques. The distributions of compatible characters all perfectly define the same group (i.e., show no homoplasy). This approach may appear indistinguishable from "cladistics" as we have described it. However, the following not-so-subtle differences as described by Farris and Kluge (1979) clarify that the two approaches are quite different.

First, for any given set of taxa there may be many maximal cliques. This would not be a problem, but there are no agreed upon criteria by which to choose among these cliques. *Second,* once the maximal clique is formed on the basis of certain character information, it is not clear what the fate of the remaining character information should be. Therefore, a large percentage of the available characters might never be involved in group formation. *Third,* the greater the homoplasy, the more limited the support for any given clique because all characters must be unique to the group they define.

In spite of the favorable comparisons of compatibility analysis with cladistics on the part of proponents of the former, clique techniques have received only limited methodological refinement and fewer applications. It would seem that the serious flaws of the method are in no way overcome by the would-be benefits.

there is no widely accepted formula or measure that can be applied for the concept of concordance.

Measures of Synapomorphy

Measures of synapomorphy are more recent than the consistency index. Farris (1988, 1989; see also Archie, 1989) used the term *retention index* for a mea-

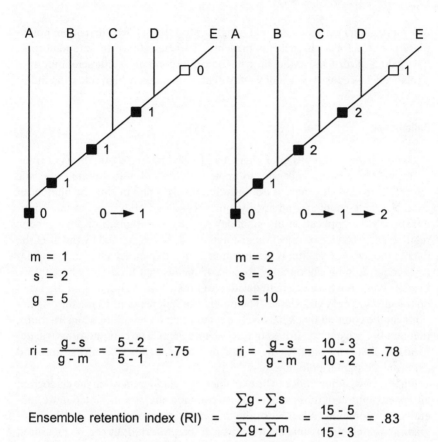

$$\text{ri} = \frac{g-s}{g-m} = \frac{5-2}{5-1} = .75 \qquad \text{ri} = \frac{g-s}{g-m} = \frac{10-3}{10-2} = .78$$

$$\text{Ensemble retention index (RI)} = \frac{\sum g - \sum s}{\sum g - \sum m} = \frac{15-5}{15-3} = .83$$

Fig. 6.2. Computing retention indices. Using the same characters and cladograms as in Fig. 6.1, we can compute the retention index, *ri*, once we know the maximum possible homoplasy, *g*, for a character on a cladogram. The values, *ri*, for the individual characters, and the ensemble retention index, RI for all characters, are computed as shown.

sure that determines the *fraction of potential synapomorphy retained as synapomorphy on a cladogram.* The retention index (ri) for a single character is computed as follows (Fig. 6.2):

$$\text{ri} = \frac{g-s}{g-m}$$

where *g* is the maximum number of steps possible for a character, *s* is the observed number of steps, and *m* is the minimum number of steps. The *ensemble retention index* (RI) can also be computed for all characters on a cladogram in the same manner as for the ensemble consistency index. A value of 1 indicates a character that is completely consistent with a cladogram, whereas smaller values approaching zero indicate that a higher proportion of the maximum homo-plasy possible for a character is present. Characters that are autapomorphic have values of zero, indicating no synapomorphy content. A character may have a relatively low consistency index but still have a relatively high retention index.

Optimization

Optimization is the fitting of characters to cladograms on the basis of some criterion. The concept is central to implementation of a quantitative cladistic approach. In cladistic analysis optimization involves minimizing the number of character-state changes — ad hoc hypotheses — on a cladogram. As such, opti-mization is the application of parsimony to the establishment of phylogenetic relationships. Choice or rejection of a hypothesis is determined by the fit of the data to the hypothesis, that is, by the number of independent character origina-tions required on a cladogram. "Required" means that for a given hypothesis, the data could not have originated with fewer than X steps, but no limit is placed on the number of "extra steps." The number of independent character origina-tions can be counted by exhaustively enumerating all possible reconstructions; that number is fixed, regardless of how we find them. The computations will de-pend only on (1) the data, (2) the "costs" of character-state transformations, and (3) the cladogram topology.

On this view, similarities can be explained by a cladogram when the cladogram allows attribution of those similarities to common ancestry or, stated more gen-erally, to a common cause. Fit of the data determines exclusively which clado-gram(s) among all possible cladograms is to be preferred; fit can be measured only through optimization.

Optimization methods are *algorithms,* a series of steps in a process, that ulti-mately allow for determination of which hypotheses best describe the data un-der a given criterion.

Optimality Criteria

Parsimony was apparently first used as an optimality criterion in phylogenetic analysis by Edwards and Cavalli-Sforza (1964) and Camin and Sokal (1965). Several variants have been proposed as representing the most realistic "theory" of character evolution. The Camin-Sokal theory of character evolution treated change as irreversible. The Dollo theory, as described by Farris (1977), treated all apomorphic states as unique, with homoplasy being accounted for only by re-versal. These approaches place definite restrictions on the possibilities for char-acter transformation. Neither theory has received widespread acceptance or ap-plication. Subsequently, the term *parsimony* has been used almost exclusively in reference to two approaches to minimizing the number of state changes on a cladogram.

Additive (Farris) optimization. Kluge and Farris (1969) and Farris (1970) employed the so-called Wagner ground-plan method of analysis, the name deriving from the American botanist W. H. Wagner, who first described the general approach. Under this criterion, character state changes are *additive* (ordered), but reversal and origination are not restricted. *Additivity assigns additional steps (costs) if changes on the cladogram do not adhere to the state-to-state order of multistate characters as defined in the data matrix* (Fig. 6.3). Optimization for additive characters can be accomplished with the algorithms published by Farris (1970), Swofford and Maddison (1987), and Goloboff (1993). The general procedure is widely known as "Farris optimization."

Nonadditive (Fitch) optimization. Fitch (1971) proposed the use of *non-additive* (unordered) transformations for the analysis of amino acid and DNA sequence data. *Non-additivity allows any state-to-state transformation — such as nucleotide for nucleotide, amino acid for amino acid — without the imposition of additional steps (costs)* (Fig. 6.4). Optimization under this "theory" of character evolution can be performed with algorithms of the type published by Fitch (1971). The method is frequently referred to as "Fitch optimization."

Alternative algorithms for optimization. If one wishes to impose transformation costs different from those adopted in the additive and non-additive approaches, Farris or Fitch optimization cannot be used to find the optimal states. In such cases, it is necessary to use algorithms of the type first developed by Sankoff and coauthors (Sankoff and Rousseau, 1975; Sankoff and Cedergren, 1983). These algorithms (dubbed "generalized parsimony" or "step-matrix" by Swofford et al., 1996) consist of enumerating all possible combinations of state assignments for every node in the tree and its surrounding nodes, calculating partial costs in every case, and choosing the best result. This is, in some sense, a "brute force" approach. Using these algorithms, any set of "costs" can be assigned a priori to the different transformations. This approach readily allows for implementation of any "model" of nucleotide substitution, for example, something not possible under non-additive (Fitch) optimization.

Optimization of characters, then, is not the application of different methods, but rather represents a choice among different rationales for assigning transformation costs between states. Ideally, such a choice should be made on the basis of evidence, as argued by Lipscomb (1992) for treating characters as additive when evidence for transformation exists in the form of observed similarity. Once the choice is made, the algorithms developed by Camin-Sokal, Farris, Fitch, and

Character	1	2	3	4	5	6	7
TAXON X	0	0	0	0	0	0	0
TAXON A	1	1	0	0	0	1	1
TAXON B	1	1	0	0	0	2	2
TAXON C	0	0	1	1	0	3	3
TAXON D	0	0	1	1	1	4	3
TAXON E	0	0	1	1	1	5	3

2 trees: length 15 steps, CI = 86 RI = 85

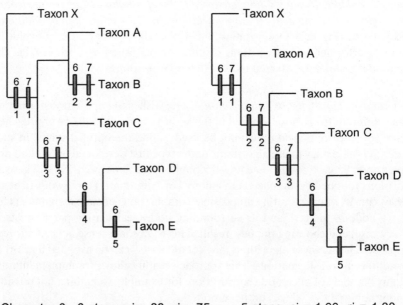

(a)   (b)

Character 6: 6 steps, ci = 83 ri = 75   5 steps, ci = 1.00 ri = 1.00
Character 7: 4 steps, ci = 75 ri = 66   3 steps, ci = 1.00 ri = 1.00

Fig. 6.3. Additive (Farris) optimization of multistate characters produces two most parsimonious cladograms with a length of 15 steps for the data matrix shown in the figure. Cladogram *a* shows the optimization of character 6, a 5-step character, with a total of 6 steps and character 7, a 3-step character, with a total of 4 steps. Notice that because the character coding is not concordant with the most parsimonious tree for the data, additive coding has added one extra step for each character, but nonetheless each contains substantial synapomorphy content. Cladogram *b* shows the optimization of both characters 6 and 7 with the minimum number of steps, the homoplasy in the data occurring in the two-state characters.

Character	1	2	3	4	5	6	7
TAXON X	0	0	0	0	0	0	0
TAXON A	1	1	0	0	0	1	1
TAXON B	1	1	0	0	0	2	2
TAXON C	0	0	1	1	0	3	3
TAXON D	0	0	1	1	1	4	3
TAXON E	0	0	1	1	1	5	3

1 tree: length 13 steps, CI = 100 RI = 100

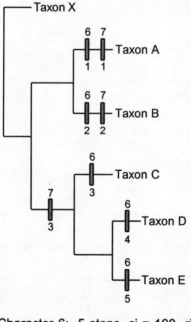

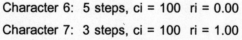

Character 6: 5 steps, ci = 100 ri = 0.00

Character 7: 3 steps, ci = 100 ri = 1.00

Fig. 6.4. Nonadditive (Fitch) optimization of multistate characters produces a single, most parsimonious cladogram with a length of 13 steps, using the same data as in Fig. 6.3, with characters 6 and 7 fitting the cladogram perfectly. All states for character 6 are autapomorphic and uninformative with regard to groupings; therefore, the retention index (ri) is 0.00. State 3 of character 7 is synapomophic for taxa 3, 4, and 5. Comparing the results of non-additive optimization with additive optimization (Fig. 6.3), it can be seen that the non-additive tree is shorter, but that information on grouping is lost in favor of treating states as unique to terminal taxa.

Sidebar 8
Metricity and the Triangle Inequality

The use of computer algorithms requires a mathematical model upon which tree construction can be based. The models used are metrics which fit the triangle inequality.

The metric used in the cladistic analysis of discrete character data is the Manhattan which is based on a *path-length model.* Where, d equals the distance between two taxa, and i, j, and k are taxa, then:

$$d(i, j) \leq d(i, k) + d(j, k)$$

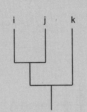

As illustrated with this cladogram, the distances between taxa i and k and between taxa j and k may be different, but their sum must be greater than the distance between taxa i and j.

Phenetic approaches, on the other hand, have employed an ultrametric model, using clustering-levels which are Euclidean distances. A measure is ultrametric when:

$$d(i, j) \leq \max[d(i, k), d(j, k)]$$

The important difference between the path-length distance and the Euclidean distance is that the former allows for variable rate of change, whereas the latter implies uniform rate of change. Indeed, it is claimed in many phenetic writings that classifications must be based on an ultrametric because this is the only way that information on "levels" in the classification can be correctly stored (Farris, 1982; see also discussion in Carpenter, 1990).

If we name the "internal nodes" in the phenogram as a and b:

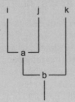

it is easy to see that the requirement that the distance between taxa i and k equal the distance between taxa j and k can be fulfilled only if the amount of evolution from a to i is the same as that from a to j and the amount of evolution from b to k be the same as that from b to i and from b to j. When some of the branches are "longer," the inequality cannot be satisfied for all triplets. This happens, obviously, when there are unequal rates of evolution.

Consider a tree where branch lengths (not levels) indicate amount of difference (with numbers by the branches indicating branch lengths):

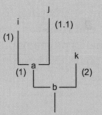

Note that the branch aj is slightly longer than ai. But then, not all three instances of the ultrametric inequality are satisfied:

$$d_{ij} \leq \max{(d_{ik}, d_{kj})} \rightarrow 2.1 \leq \max{(3, 3.1)} \qquad \text{TRUE}$$
$$d_{ik} \leq \max{(d_{ij}, d_{kj})} \rightarrow \ \ 4 \ \leq \max{(2.1, 4)} \qquad \text{TRUE}$$
$$d_{jk} \leq \max{(d_{ik}, d_{ij})} \rightarrow 4.1 \leq \max{(4, 2.1)} \qquad \text{NOT TRUE}$$

In such a case, where the observed distances (the actual amount of difference) between taxa are unequal, it will not be possible to produce a tree diagram with the observed distances perfectly retrievable from the distance levels. But path-length distances as used in cladistics will produce a perfect fit. In fact, as shown by Farris (1979), whenever the "levels" model produces perfect fit, so will the "path-length" model.

Sankoff are the tools used to find the best state assignments given those transformation costs.

The various optimality criteria can be summarized as follows:

Additive (ordered; Farris)	Change costs increase with alteration of state-to-state order.
Non-additive (unordered; Fitch)	Change costs are equal for any possible state-to-state change.

Camin-Sokal	Reversals are of infinitely high cost.
Dollo	Multiple origins of same character are of infinitely high cost.
Sankoff	Costs for any state-to-state change are assignable on a differential basis.

Storing Trees

Trees cannot be handled by the computer as branching diagrams. Thus they are stored in *parenthetical notation* whereby nested sets of parentheses are used to indicate the inclusive nature of groupings. Examples are shown in Fig. 6.5. Note that the number of left- and right-hand parentheses is "balanced" in the examples. Computer phylogenetics programs produce notation that is balanced, although they may be able to accurately interpret trees for which not all groupings have balanced parentheses.

Finding Optimal Trees

Exact parsimonious solutions for small data sets can be discovered by enumerating all possible trees. There will, however, never be verified exact solutions for larger data sets because there is no known mathematical solution for the type of problem being solved (see Day, Johnson, and Sankoff, 1986), and because no matter how much computing power is available, there are more possible trees than can ever be examined (Fig. 6.6).

Most currently used phylogenetic algorithms adhere in principle to the method originally proposed by Kluge and Farris (1969) and Farris (1970) for calculating Wagner trees. In order to improve the success rate for finding optimal solutions, many variant implementations of Farris's original approach have subsequently been developed, increasing immensely the speed and effectiveness with which optimal trees can be found.

The Wagner method of Farris produces — as do some other methods — "unrooted trees," often referred to as *networks*. Because the parsimony criterion does not require a priori considerations of polarity or synapomorphy, it is not necessary to root the scheme of interrelationships in order to optimize the character states for all internal nodes on the tree. Thus, determining the optimal set of character states for the internal nodes under a given optimality criterion is the core problem of numerical cladistics. The states for the terminal taxa are already known!

A Wagner tree (network) is formed by the sequential addition of taxa. Because different addition points produce trees of different lengths, taxa are added

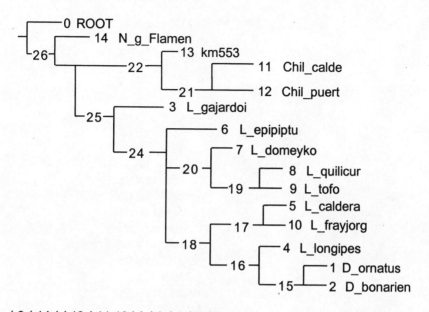

$$(0 (14 ((13 (11\ 12)) (3 (6 (7 (8\ 9)) ((5\ 10) (4 (1\ 2))))))))$$

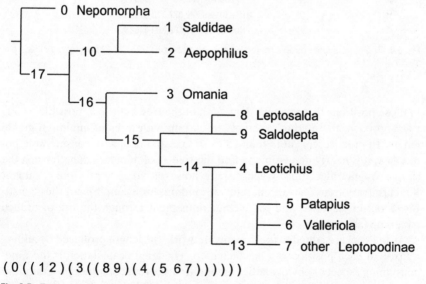

$$(0 ((1\ 2) (3 ((8\ 9) (4 (5\ 6\ 7))))))$$

Fig. 6.5. Examples of parenthetical notation and corresponding cladograms.

Rooted Bifurcating Trees

1	1
2	1
3	3
4	15
5	105
6	945
7	10,395
8	135,135
9	2,027,025
10	34,459,425
11	654,729,075
12	13,749,310,575
13	316,234,143,225
14	7,905,853,580,625
15	213,458,046,676,875
16	6,190,283,353,629,375
17	191,898,783,962,510,625
18	6,332,659,870,762,850,625
19	221,643,095,476,699,771,875
20	8,200,794,532,637,891,559,375
21	319,830,986,772,877,770,815,625
22	13,113,070,457,687,988,603,440,625

Fig. 6.6. The numbers of possible rooted bifurcating trees for 1 through 22 taxa (after Felsenstein, 1978).

in those positions that increase the length of the tree as little as possible. Using this approach, the method attempts to end up with trees of minimum global length. In practice, for more than 15 to 20 taxa and data with considerable homoplasy, it is nearly impossible to find the tree(s) of minimum length using the simple Wagner algorithm. Wagner trees, however, are generally *much* shorter than arbitrary trees (those generated at random, for example) and thus constitute a better starting point for further refinement through the use of branch swapping (see below).

As the taxa are added to the growing network, the length produced by adding a taxon at each position can be calculated. The length calculation is the time-consuming aspect of phylogenetic computation; current computer programs use fast indirect methods to calculate length.

Note that different addition sequences of the taxa may lead to different trees. Consider the data matrix in Fig. 6.7, where taxon X represents the outgroup.

Forming a tree with a simple Wagner algorithm, which adds the taxa in the order in which they appear in the matrix, produces an incorrect tree (Fig. 6.7a). That is because the first three characters determine the unique tree (X ((A B) (C D)), with the last two characters appearing as homoplastic. However, if only X, A, and B have been placed in the tree, when C is to be added the characters that join C and D will appear as uninformative (that is, autapomorphic, requiring the same number of steps for any possible placement of C), since D is still not placed in the tree. If the taxa are instead added as (XAC),D,B the shortest tree is always obtained (Fig. 6.7b). In other words, the evidence that the Wagner method considers at each step is partial; thus, a taxon position optimal for all of the evidence might not appear as optimal for only part of it (Goloboff, 1998).

Phylogenetic Algorithms

There are now scores of phylogenetic algorithms available for finding trees under the optimality criteria discussed above. The existence of variant approaches to solving essentially the same problem derives not from a plethora of optimality criteria, but rather from the fact that finding exact solutions is unfeasible for all but relatively small data sets (i.e., more than 15–20 taxa). Finding the best trees for large data sets, particularly those with substantial homoplasy, is *very* difficult. Thus, as you might expect, there are many ways to implement solutions.

The following descriptions of algorithmic approaches to solving phylogenetic problems are general. More elaborate sources are available for those who are interested (e.g., Goloboff, 1994, 1996; Swofford et al., 1996).

Algorithms can be divided into two classes on the basis of the solutions they produce: *exact* and *heuristic* (approximate solution).

Exact-Solution Algorithms

Exact solution algorithms are of two types. First, *exhaustive search* algorithms examine every possible cladogram for a given set of taxa; the maximum number of allowable taxa in the data set will probably be 12, as a consequence. Second, *branch-and-bound* (implicit enumeration) algorithms perform exhaustive searches within limits (bounds) established during the process of searching for minimum-length cladograms, rather than examining all possible cladograms.

A brief explanation may be helpful for understanding how the branch and bound approach works. Generating all possible trees would require that the first three taxa be joined, then a fourth taxon be added in the 3 available positions, the fifth be added in the 5 available positions, and so on. The length of the tree would be calculated for all possible topologies as each new taxon is added to

Character	1	2	3	4	5
TAXON X	0	0	0	0	0
TAXON A	1	0	0	0	0
TAXON B	1	0	0	1	1
TAXON C	0	1	1	1	1
TAXON D	0	1	1	0	0

(a)   Simple Wagner algorithm = 8 steps

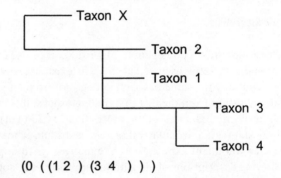

(0 ((1 2) (3 4)))

(b)   Randomized addition sequence = 7 steps

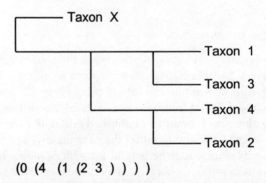

(0 (4 (1 (2 3))))

Fig. 6.7. Matrix of five characters and five taxa, with taxon X as the outgroup, indicating how using a simple Wagner algorithm (a) may fail to find a correct solution, whereas randomizing the taxon addition sequence (b) achieves the correct solution (see text for additional explanation).

the ever-growing tree. Because the length of a tree can never decrease as more taxa are added, if the length of any partial tree exceeds that of the best complete tree(s) found so far (say 4 taxa produced a tree longer than 5), it is unnecessary to continue with the addition of more taxa on the topology represented by that longer partial tree. This approach allows for the rejection of many trees by examining/evaluating only a few partial trees. For example, if the lengths for two positions for the fourth taxon exceed the length for the best complete tree yet found, this will allow for the rejection of all the cladograms derivable from those two topologies: two-thirds of all possible topologies can be rejected just by optimizing the subtrees formed by the three possible addition points for the fourth taxon. In this way, the method can reduce the number of cladograms that need to be evaluated, thereby increasing the number of taxa to at least 18 for which a guaranteed optimal solution can be found (see also discussion in Swofford et al., 1996).

Heuristic (Approximate Solution) Algorithms

Beyond 18 or so taxa we must rely on *heuristic (approximate solution) algorithms.* The simplest of these is the *Wagner algorithm,* which, as discussed above, creates trees by successively adding taxa to a growing network. In its most basic implementations, the Wagner algorithm is very fast but does not produce most parsimonious trees for data that are not highly congruent, as illustrated in the example in Fig. 6.7a. The results can often be improved without greatly increasing the time needed to perform the calculations through the use of random taxon-addition sequences. Trees produced by the Wagner algorithm may be used as "starting trees" for branch swapping.

The remaining algorithms modify preexisting trees via branch swapping, in their attempts to identify the most parsimonious tree(s). *Branch swapping* is a process in which branches on a tree are moved in an attempt to find more promising topologies — that is, shorter trees. Two approaches are commonly applied: *tree bisection-reconnection* and *subtree pruning-regrafting* (see Swofford et al., 1996). The latter approach examines more trees, and therefore takes longer, but it is also likely to produce better results. The efficiency and effectiveness with which branch swapping finds shorter trees will depend on how close the input tree is to minimum length, the effectiveness of the branch-swapping algorithm itself, the amount of homoplasy in the data, other aspects of data structure, and — of course — the speed of the computer. Detailed discussions of algorithmic approaches can be found in Goloboff (1996), Swofford et al. (1996), and Gladstein (1997).

Because branch swapping does not produce an exact result, one might wish to have a method for determining when a search has found an optimal tree. This can be done as follows: If the swapper performs 20 replications (distinctive addition sequences), and all of them find the same two trees, then stop. If the swapper performs 100 replications, and 2 replications find trees of 100 steps while the other 98 replications find trees of different (greater) lengths and topologies, then additional replications are required until a more consistent result of 100 or fewer steps is achieved.

Most available phylogenetics program packages contain algorithms that produce comparable results. They may not produce them in comparable times, however, for the reasons noted above. For a review of dated comparisons of efficacy and speed of execution of phylogenetic programs, see Platnick (1989).

Character Weighting

Quantitative cladistic analysis lends itself to weighting, some optimality criteria involve weighting, and most computer phylogenetics programs make provision for weighting. Thus, it will be helpful to gain some perspective on this subject, which has caused much controversy and considerable confusion.

Traditionally, character weighting has been treated as being of two types: *a priori,* the weights being applied without regard to the behavior of the characters, and *a posteriori,* the weights being related to the behavior of the characters themselves. These approaches might also be labeled *subjective* and *objective* weighting.

A Priori (Subjective) Weighting

Weighting criteria such as relative morphological complexity, functional importance, adaptive significance, and probability of state inheritance in DNA sequence data are inherently subjective. To understand why, we might consider the result of giving complexity higher weight than simplicity. In such a case, loss might be weighted low because it represents a reduction in complexity. Nonetheless, loss characters can be excellent group-defining attributes, as, for example, the absence of limb elements in snakes. Experience suggests not only that many apparently "simple" characters define groups perfectly but also that complexity is subject to varied conceptions, depending on the observer.

An unvarnished approach to subjective character evaluation comes from Guttmann (1977:646):

> Only those features and characters whose functions are known, and for which the value of the adaptational changes can be assessed, can be utilized in phylogenetic construction. A phylogenetic theory can neither be constructed by comparison of morphological configurations nor by character analysis.

That this view suggests a priorism would seem clear on its face.

Such a priori views are not restricted to discussions of morphological characters. A clearly stated vision concerning the primacy of molecular data, in the absence of support for the argument, was expressed by Sorensen et al. (1995):

> phylogenetic reconstruction using nucleotide sequencing is thought to be superior
> to, and definitely more objective than, that based upon morphology. . . . This is be-
> cause, in general, nucleotide substitutions are random, non-selective events, as op-
> posed to trying to determine how to code and weight morphological characters,
> which are de facto a result of selection.

If nucleotide substitutions were truly random, we might expect the results of
their analysis to be without hierarchic structure. Yet, hierarchic structure is exactly what Sorensen and all others performing phylogenetic analysis seek. As is
emphasized elsewhere in this text, there is strong empirical support for the view
that congruence exists between morphological and sequence data. When these
two data types are analyzed simultaneously, the ratio of phylogenetic signal to
noise — that is, the ability to recover hierarchy from the data — is often improved.
Thus, the subjective statements of Sorensen et al. (1995) and others concerning the superiority of sequence data would seem to represent little more than
opinion.

The a priori nature of using the probability of a given nucleotide being inher-
ited at a given site can be understood through the realization that the approach
requires us to assume that all changes fit some model chosen prior to performing cladistic analysis. Goloboff (1993) and Wenzel (1997) have described such
approaches as producing phylogenetic results as good as our prior assessments
of the reliability of characters. Only under circumstances where the model was
accurate for all substitutions could the weights be justified. Such information
can hardly be available before the analysis is conducted.

It would seem clear that if the quoted viewpoints by Guttmann and Sorensen
et al. were accepted, most of the data ever used in systematics could not be admitted as valid. The subjective nature of such views is emphasized in the contrast
between them, one suggesting that adaptive significance is of paramount importance, the other suggesting that valid character data must be nonadaptive.

Such arguments invariably rely on a priori appeals to the value of characters.
The "goodness" of characters as indicators of relationship cannot be judged a
priori, however. Instead, it can only be judged objectively with respect to congruence among characters themselves. Otherwise, ad hoc hypotheses of adaptive significance and selective neutrality could be invoked ad infinitum. Thus,
weighting strategies that apply "objective" criteria are the only ones to which we
will devote further attention.

A Posteriori (Objective) Weighting

Two remaining approaches discussed by Goloboff (1993) in his review of
weighting approaches can be labeled as "objective."

Weighting based on character compatibility. First among these is weighting based on character compatibilities. Goloboff noted that a compatibility-derived weighting scheme is not based on the degree of correlation between characters and trees (an obvious violation of all parsimony-based optimality criteria) because characters with fewer incompatibilities may be more homoplastic. The most obvious implementation of this particular approach is "compatibility analysis" itself (see Sidebar 7), as first proposed by Le Quesne (1969; see also Sharkey, 1989). Compatibility analysis requires that all characters used to construct a phylogenetic scheme be perfectly consistent; that is, all must be in absolute agreement concerning the scheme they define. Such an approach could be described as an all-or-none weighting strategy. Unlike the consistency-based weighting approaches described below, in compatibility analysis it is only the "compatible characters" that define the scheme because they are the only ones with a weighting value. All other character data are given a weight of zero. As the level of homoplasy increases, the level of compatibility decreases, such that the method explains fewer and fewer of the original observations.

Weighting based on character consistency. Second, Goloboff discussed weighting based on consistency, the more homoplasious characters being those that receive lower weights. Probably best known among this type of approach is *successive approximations weighting,* first proposed by Farris (1969) as a method of evaluating the strength of characters as they define a given tree. The method functions as follows:

1. Compute the most parsimonious tree(s) for a data set with equally weighted characters.
2. Weight the characters in the data set relative to their agreement with the cladogram(s) they define (e.g., by the value of their consistency index).
3. Recompute the tree(s) based on the characters with their newly assigned weights.
4. Continue the process iteratively until the weights become stable.

This approach assigns equal weights to the characters for purposes of computing the "input" (initial) tree(s). Successive weighting then determines the values (weights) that *should* be assigned to the characters on the tree(s) that the equally weighted characters defined.

If multistate characters are included in a result subjected to successive approximations weighting, they should be recoded in additive binary format. Under this approach, consistencies for all state changes are evaluated independently, and the weighting function will therefore be applied on a comparable basis for all information on character transformation (Farris, 1969; Carpenter, 1988).

Successive approximations weighting was originally justified by Farris as being consistent with the logic of parsimony because the weights are assigned based on the consistency of the characters. The procedure does not, however, implement an *optimality criterion* by which the fit of data to a tree can be measured directly. Successive approximations weighting can only be applied to data on the basis of a cladogram produced by prior analysis with those data (see further discussion in Chapter 7).

The above discussion of weighting based on character consistency assumes at the outset that all characters are given equal weights for the purposes of phylogenetic analysis. It offers no justification for treating characters as having different weights prior to or during phylogenetic analysis. The validity of the assumption of prior equal weights can be scrutinized through the application of successive approximations weighting, but the cladograms produced by that method are often not changed, only "filtered." The applicability of the equal weights assumption has, however, been challenged by Goloboff (1993) on the grounds that all characters do not contribute equally in determining cladogram topology and therefore should not be given equal weights for the purpose of computing cladograms.

Implied weighting and fittest trees. Goloboff (1993) extended the logic of weighting on the basis of consistency to an approach that evaluates cladograms according to the "implied weights" of the characters. Goloboff (1993:85) justified his approach by noting that a tree that is shortest under the weights it implies is *self-consistent,* resolving conflict among characters in favor of those that have less homoplasy and therefore provide stronger phylogenetic signal. Conversely, a tree that is not shortest under the weights the characters imply is *self-contradictory* because conflicts are resolved in favor of those characters the tree is telling us not to trust — that is, the ones with the greatest amount of homoplasy.

The idea of self-consistency also applies to the results of successive approximations weighting, where the iterative weighting procedure is repeated until a stable result is achieved. That result is self-consistent, with the weights of characters leading to a preference for the tree that implies them. Although successive approximations weighting has generally been applied in cases where a parsimony analysis produces multiple trees (see Chapter 7), as pointed out by Goloboff (1993), self-consistency of a single tree might just as reasonably be evaluated through the application of successive approximations weighting. The starting point for trees subjected to successive approximations weighting is based on calculations using characters given prior equal weights, whereas when searching for fittest trees under the implied weights procedure, weights are assigned during the tree searching process.

The weight of a character, then, can be directly related to the fit of that character to the tree, as measured — for example — by the consistency index. Under Goloboff's approach, the value to be maximized is *total fit,* the sum of the fits of all characters to a tree. In order to relate fit to homoplasy, Goloboff's implied

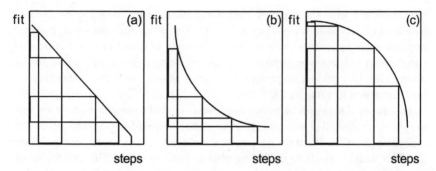

Fig. 6.8. Types of weighting functions. Under a linear function (a), the same step difference is equally important regardless of the amount of homoplasy (i.e., fit and homoplasy have a linear relationship); therefore, the slope of the line changes with the weights of the characters, but not with respect to homoplasy. Under the concave fitting function (b), the value of a character decreases relatively rapidly as homoplasy increases from its minimum value and decreases slowly as homoplasy approaches its maximum value, as seen from the slope of the curve. Under the convex fitting function (c), the value of a character decreases relatively slowly as homoplasy increases from its minimum value and increases rapidly as homoplasy approaches its maximum value, as seen from the slope of the curve (adapted from Goloboff, 1993).

weights approach requires the application of a weighting function, as does successive approximations weighting. If that function is linear, it treats the same step difference as equally important regardless of the amount of homoplasy (Fig. 6.8a), which would be the case if all characters were given prior equal weights. This might be called the traditional approach under parsimony. If all characters were given higher prior weights, the slope of the line would become steeper, but the weighting function would not differ for characters with different amounts of homoplasy.

Goloboff argued for the fittest trees approach, as Farris (1969) had for successive approximations weighting, that a concave weighting function (Fig. 6.8b) appeared to be the most desirable because it gives the highest weights to characters with maximum group-defining value. Such a function gives distinctly and conspicuously lower weights to characters with limited group-forming value — that is, those with the most homoplasy. Under this criterion, the addition of an extra step to an already highly homoplastic character changes the value of the character much less than adding an extra step to a character with no homoplasy at all.

The remaining possibility, a convex weighting function (Fig. 6.8c), produces what Goloboff referred to as the "absurd" possibility of weighting characters with more homoplasy more heavily than those that are perfectly consistent with the tree they define.

The consistency index is a concave function and therefore provides the initial basis for the weighting function. The concavity of this curve, and therefore strength of the function, may be greater than one wishes, however. Thus, a *constant of concavity* can be added to change the severity of the weighting function.

A value of 0 for the constant of concavity maintains the concavity of the consistency index, whereas higher values reduce the strength of the weighting function. As pointed out by Goloboff, there are no empirical studies indicating exactly what values for the constant of concavity will be most suitable for data sets with certain characteristics.

The "fittest trees" method of Goloboff utilizes either Farris or Fitch optimization to calculate state assignments and measure the homoplasy for a given (single) character, exactly as for finding shortest trees. The fittest tree(s) may not be the shortest tree(s) because fit maximizes the sum of the average unit consistency index (ci), as shown in the following formula:

$$\sum ci = \sum \left(\frac{m}{s}\right)$$

whereas the shortest tree maximizes the ensemble consistency index (CI):

$$CI = \frac{\sum m}{\sum s}$$

As before, m is the minimum number of steps for a character on the cladogram, and s is the actual number of steps for the character on the cladogram. Under Goloboff's method, weights are assigned during the tree searching and comparison process. Under successive approximations weighting, weights are based on the prior analysis, and reanalysis is performed iteratively until self-consistency is achieved.

Islands of Trees and Solutions for Very Large Data Sets

The use of branch-swapping algorithms may produce less-than-optimal results for some data sets (Maddison, 1991) because once an algorithm has found a promising topology — a local optimum or "island" — it may neglect other equally promising topologies simply by virtue of its design. Until recently, the most effective solutions to this problem have been (1) the use of different starting points (i.e., a tree slightly longer than the shortest tree found without the use of exhaustive branch swapping) from which to begin branch swapping, and (2) using more exhaustive approaches to branch swapping. The former solution has been the more widely employed. Because of the island phenomenon, it is possible that starting to swap on a tree of 100 steps will result in being stuck on a suboptimal "island," whereas starting to swap on a randomly generated tree of 1,000 steps might ultimately produce shorter trees. The result in the latter case, however, would almost certainly take much longer to produce.

When data sets become very large, say more than 200 taxa, the total number of possible solutions is immense and thus recovery of any solution — let alone effectively surmounting the island phenomenon — might be viewed as problematic. One view of the computational state of affairs as of 1998 was offered by Soltis et al. (1998). These authors showed that combining data for different DNA regions increased the amount of "signal" in the data and that solutions could therefore be found in "days" rather than the weeks or months that had previously been required with the software they were using. The speed with which most phylogenetic algorithms arrive at solutions is directly related to strength of phylogenetic signal; this property has long been known, although it was not mentioned by Soltis et al. What those authors found, however, was that whereas individual DNA-sequence data sets (regions) with little phylogenetic signal often precluded achieving a computational result, combining these data sets increased the phylogenetic signal, computational times decreased dramatically, and solutions were effected.

Some authors have suggested adopting inferior methods of analysis in cases where available parsimony algorithms were apparently incapable of processing the data at hand. An example is the use of "neighbor joining" (Saitou and Nei, 1987) — a technique described by Swofford et al. (1996) as lacking an optimality criterion and therefore purely algorithmic — which produces a single tree in a limited period of time. No matter that neighbor joining requires that discrete data first be converted to distance data, that the form of the tree is influenced by the order of taxon input, and that only a single tree is produced, whereas many preferred (shorter) trees may exist for the data set, as was shown by Farris et al. (1996).

One alternative approach to addressing the problem of very large data sets was proposed by Farris et al. (1996). Instead of using methods such as neighbor joining or relying on branch swapping algorithms to find phylogenetic solutions, their "Jac" program uses "jackknife" resampling procedures (see Chapter 7) and a fast parsimony algorithm to implement large numbers of computations in a relatively short period of time. The results, in the form of a consensus, include only groups that are well supported by the data and exclude cladograms containing groups lacking support. As noted by the authors, poorly supported groups cannot survive resampling and are therefore automatically eliminated from the results. The justification for the Jac approach was that the use of conventional parsimony algorithms required weeks of computational time for some data sets, and the consensus of the thousands of trees resulting from the parsimony analysis was very poorly resolved. The Jac approach produces results that include only well-supported groupings and that would be useful as predictive classifications. It also provides a relatively efficient way to find a better "starting tree" for branch swapping on very large data sets.

No matter what the benefits of using Jac, it is now clear that neither days of computing time as required in the study of Soltis et al., nor resampling procedures as utilized by Farris et al., will be necessary to produce credible solutions for relatively large data sets. Nixon (1998) has shown that the "Parsimony Ratchet," a new search method that can be easily implemented with existing phylogenetic software, has the ability to produce solutions for large data sets in a realistic time frame. The Ratchet is implemented through the following steps:

1. Generate a starting tree (e.g., a Wagner tree), followed by some level of branch swapping.
2. Randomly select a subset of characters, each of which is given additional weight (e.g., add 1 to the weight of each selected character).
3. Perform branch swapping (e.g., tree bisection-reconnection) on the current tree using the reweighted matrix, keeping only one or a few trees.
4. Set all weights for the characters to the "original" weights (typically, equal weights).
5. Perform branch swapping again on the current tree (from step 3), holding one (or few) trees.
6. Return to step 2.

Steps 2 through 6 are considered to be one iteration, and, typically, 50–200 or more iterations are performed. The number of characters to be sampled for reweighting in step 2 is determined by the user. Weighting between 5 and 25 percent of the characters provides good results in most cases.

Because the Ratchet samples many tree islands with a few trees from each island, it provides much more accurate estimates of the "true" consensus than collecting many trees from few islands. Using the Ratchet in combination with readily available microcomputer hardware and software can produce results in a few hours or days, whereas those same analyses previously required months or years for analysis, as was the case in the study of Soltis et al. (1998). Test runs indicate efficiency increases of 20–80 times over what has previously been possible. A concurrent decrease in the length of the best trees found is also often observed.

Rooting: Additional Discussion

When transformation costs are symmetrical for a given data set, different rootings of a given tree are the same; that is, the rootings are transparent to the parsimony criterion and produce trees of the same length under "standard" parsimony, or equivalent fit under Goloboff's method of implied weights. Therefore, the explanatory power of all possible rootings of a given network is the same.

To understand this concept, consider the following tree, which is depicted graphically in Fig. 6.9:

$$(0(0(1(1(1(1(00)))))))$$

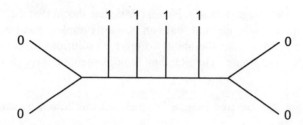

Fig. 6.9. An unrooted tree (network), depicting graphically the character distributions indicated in parenthetical notation in the text.

Which state is "apomorphic," 0 or 1? This cannot be answered. At most, one can say that "in the transformation from 0 to 1 at the base of the tree, 1 is derived, and in the transformation from 1 to 0 in the tip of the tree, 0 is derived," as shown in Fig. 6.9. The only way to determine polarity, then, is to have a tree optimized and rooted, at which time the polarities of each transformation can be read off the tree. As noted in Chapter 4, the root is usually fixed by including an outgroup established by recourse to previous, higher-level cladistic analysis (see also Rooting, Chapter 4).

Alternative Approaches to Phylogeny Reconstruction

Three-Taxon Statements

A novel approach to deriving phylogenetic results from discrete character data was proposed by Nelson and Platnick (1991) under the name "three-taxon statements." A given matrix can be coded under this method as the number of three-taxon statements the characters represent. The recoded matrix (Fig. 6.10) can then be analyzed using conventional parsimony algorithms. Nelson and Platnick indicated that the method might represent a more precise use of parsimony and, in some cases, reduce the number of most parsimonious trees. An extended discussion of the method can be found in Kitching et al. (1998).

As with any new approach, the three-taxon method has not gone uncriticized, as, for example, in a paper by Farris et al. (1995). The primary criticism of the three-taxon method by those authors was that it does not conform to the justification offered for parsimony because it does not minimize the number of ad hoc hypotheses of homoplasy, treating as independent quantities that are interrelated.

Maximum Likelihood

We now return to the subject of maximum likelihood, introduced in Chapter 3. Interest in this approach derives primarily from the work of Felsenstein

Matrix 1

Characters

Taxon	1	2	3	4	5	6
A	1	0	1	0	0	0
B	1	0	0	0	1	1
C	0	1	1	1	0	1
D	0	1	0	1	1	1

Matrix 2

Three-taxon statements

	1	2	3	4	5	6
Taxon	ab	ab	ab	ab	ab	abc
A	11	?0	11	0?	0?	000
B	11	0?	0?	?0	11	11?
C	0?	11	11	11	?0	1?1
D	?0	11	?0	11	11	?11

Original data:

(1)   (2)

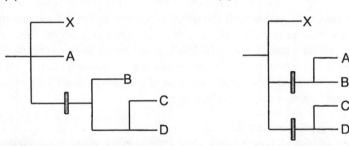

Fig. 6.10. Coding of data for the three-taxon method. Matrix 1 equals original data; matrix 2 equals the same data coded for analysis by the three-taxon method (from Nelson and Platnick, 1991). Each original character requires columns in the new matrix equivalent to the number of three-taxon statements it implies. Thus, character 1 implies C(AB) and D(AB), with the fourth taxon of each alternative coded as missing. Character 6, by contrast, implies 3 three-taxon statements, A(BC), A(BD), and Ad(CD). The original data produces two equally parsimonious cladograms (1 and 2). The three-taxon method produces only cladogram 1.

(1973, 1978; see also Swofford et al., 1996; Sober, 1988), who was of the opinion that although few prior attempts had been made to apply statistical inference procedures to the estimation of evolutionary trees, maximum likelihood (or some other statistical inference method) *must* be used to infer them. Maximum-likelihood methods have gained adherents, particularly among those involved in the analysis of DNA sequence data. The appeal of maximum likelihood would seem to derive from the presumed superiority of statistically based techniques for data analysis and the widespread acceptance of probabilistic models of sequence evolution among workers who gather and analyze such data.

Swofford et al. (1996:429), in their discussion of the likelihood approach, offered the opinion that parsimony ignores information on branch lengths when evaluating a tree, whereas maximum likelihood treats changes on long branches as more likely than those on short branches. This "important component" of the method thus explains, according to Swofford et al. (1996:430), the "consistency" of the maximum-likelihood method for cases where parsimony is inconsistent. It should not be forgotten, however, in the evaluation of this argument, that parsimony does compute branch lengths, but the lengths of branches are relevant only to the extent that they represent the minimum number of character-state changes necessary to form a most parsimonious solution. As framed by Swofford et al., the branch-length argument is therefore irrelevant and misleading as a criticism of the parsimony approach.

The Swofford et al. argument might alternatively be interpreted as meaning that maximum likelihood uses the information provided by all of the characters in terms of the length of the branch (i.e., the "evolutionary rate" along the branch), whereas parsimony treats every character independently and thus ignores the information on change rate along a given branch. This point is also irrelevant as a criticism of parsimony — or conversely as a presumed benefit of maximum likelihood — for it has been shown by Farris (1979) (one among many possible citations) that path-length distances as applied under the parsimony criterion always store information on genealogy and similarity as effectively as possible. Thus, no information would be lost in a parsimony analysis compared to a likelihood analysis, irrespective of the comments of Swofford et al.

Calculating a maximum-likelihood tree, as described by Swofford et al. (1996:431), would involve the following steps:

1. Begin with aligned sequence data.
2. Produce an unrooted tree for the data (presumably using a parsimony algorithm).
3. Root the tree arbitrarily.
4. Compute the likelihood of particular sites as the sum of the probabilities of every possible reconstruction of ancestral states given some model of base substitution that has been chosen a priori.
5. Compute the likelihood of entire tree as product of the likelihood of all sites.

Knowing the steps required to implement maximum-likelihood analysis we might conclude the following about the status of the method in the calculation of phylogenetic relationships:

First, in order to perform the calculations, one must have a model of evolution for the sequence data in question. Swofford et al. (1996) and others have discussed the nature of such models. Clearly the number of possible models is great. Those models must consider all sites, for all organisms, for all time. Yet, the criteria by which they may be judged are not clear. With regard to models,

Wenzel (1997:36) observed, as have others, that "methods [such as maximum likelihood] are self-fulfilling and do not provide independent evidence of their legitimacy. Trees built according to a given model cannot refute the model, and therefore it is clear that the model . . . is beyond testing." Parsimony-based cladistic methods, on the other hand, are founded solely on evidence. As pointed out by Siddall and Kluge (1997), even if evolution did not exist, parsimony would still be applicable.

Second, although models of change for molecular sequence data can be easily constructed, if not easily justified with regard to their reality, the same cannot be said for models of change for morphological characters. For example, on what basis does one determine the probability of change from one state of a morphological character to another — from limbs to wings — and how would one compare probabilities across characters, especially when our experience makes it abundantly clear that rates of change for different characters *are* different? The rationale for applying maximum-likelihood methods across the spectrum of character data remains unspecified. Indeed, the maximum-likelihood approach has only been advocated for — and applied to — the analysis of molecular sequence data. Thus, arguments for its superiority are unconvincing if we are searching for a method capable of analyzing all data with equal effectiveness.

At present, the complexity of obtaining maximum-likelihood solutions to problems that involve many alternative hypotheses inhibits the more general use of these techniques (Swofford et al., 1996:430).

Thus, notwithstanding the possible allure of statistical consistency, maximum-likelihood methods possess several inherent and conspicuous limitations. And, paradoxically, unlike approaches used in other fields of science where data inform theory, with maximum likelihood, theory informs data (see further discussion in Chapter 3 under Statistics, Probability, and Models as Alternatives to Parsimony).

Long-Branch Attraction

The discussion of maximum-likelihood techniques leads logically to the discussion of "long-branch attraction." This is the would-be phenomenon whereby branches with large numbers of characters evolving in parallel group together, even though the terminal taxa are not each others' closest relatives. The source of this idea comes — like maximum-likelihood as a phylogenetic technique — from the work of Felsenstein (1978) in his attempts to show that parsimony methods can be "statistically inconsistent" under certain evolutionary models. Although Felsenstein believed that the parsimony approach — in the sense that it is generally applied in this book — could produce erroneous results, he also stated as an alternative that "the conditions [as applied by him] which must hold in order to have lack of consistency may be regarded as so extreme that . . . [his] result may be taken to be a validation of parsimony" (Felsenstein, 1978:402).

There was, and still is, some question of whether "long-branch attraction" actually occurs or whether it has been accepted as occurring only by assuming that real data fit Felsenstein's models.

Good empirically based — as opposed to simulation based — discussions of long-branch attraction are those of Whiting et al. (1997), Whiting (1998), and Siddal and Whiting (1999). An analysis of phylogenetic relationships among the orders of holometabolous insects by Whiting et al. (1997) consistently produced the group Halteria (Diptera + Strepsiptera), among others. These results were based on the analysis of molecular and morphological data, comprising the largest sample of taxa and characters ever used for a higher-level analysis of relationships within the group.

The Strepsiptera, a group of internal parasites of other insects, have novel morphology and appear on what might be termed a long branch. Their phylogenetic position within the insects has long been a subject of discussion and controversy. On the basis of morphology alone, the Strepsiptera had frequently been treated as related to the Coleoptera. The grouping of Diptera + Strepsiptera, as originally proposed by Whiting and Wheeler (1994) (Fig. 6.11), was summarily rejected by Carmean and Crespi (1995) as being the result of "long-branch attraction" even when appearing as a result in their own analysis, which was based on a small sample of characters (comprising only DNA sequences) and taxa.

In defense of the Halteria as a natural group, Whiting et al. (1997:38) made the following observations:

> A point that is often missed in long-branch attraction discussions is that relative rates of substitution are influenced by phylogeny, and thus we should not be surprised to find cases of sister groups that have high rates of nucleotide substitution; shared elevated rates could indeed be evidence of shared history. The supposition that clades best supported by character data are the ones we should be most suspicious of has the strange result of entailing an inverse relationship between phylogenetic evidence and phylogenetic conclusions. The large amount of molecular evidence supporting the monophyly of the Halteria and Amphiesmenoptera can be taken at face value as indicative of well-supported sister-group relationships and not as foibles of the data or analytical method.

The assertions of Carmean and Crespi were independently investigated by Huelsenbeck (1997). He pointed out that no empirical examples of long-branch attraction existed in the literature, but he nonetheless assumed that the phenomenon may still exist as an artifact of parsimony methods because they can "converge to an incorrect genealogical tree *as more data are added,* when the assumptions of the method are severely violated" (Huelsenbeck, 1997:69) (italics added). Huelsenbeck made no mention of what assumptions are violated in causing parsimony to arrive at "incorrect" answers. His study involved the use

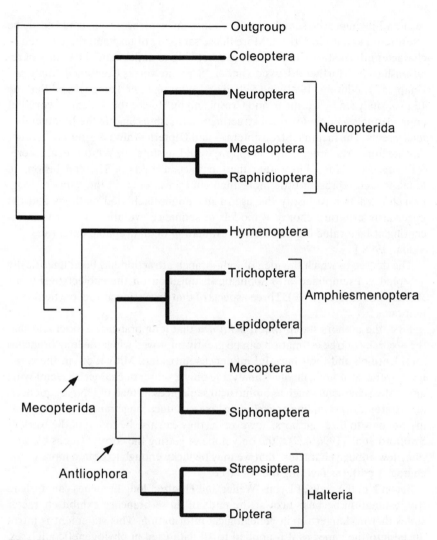

Fig. 6.11. The groupings of holometabolous insects recognized by Whiting et al. (1997). Heavy lines subtend strongly supported groups; thinner solid lines subtend relatively well-supported groups; dashed lines subtend groups with questionable support.

of simulated data analyzed under a variety of models of DNA evolution, an example with no more empirical content than that of Felsenstein (1978).

The real significance of Huelsenbeck's assertions may revolve around the issue of addition of more data (i.e., improved sampling). Huelsenbeck consistently retrieved Diptera + Strepsiptera as did Whiting et al., even though he analyzed the much smaller data set of Carmean and Crespi. To believe, as Huelsenbeck apparently does, that sampling fewer taxa and characters will produce a superior result, while improved sampling will produce an incorrect result, could easily be interpreted as anti-empiricist. More dense sampling of taxa and the addition of character information can do nothing if not help to understand the reality of re-lationships (see further discussion under "Separate versus Combined Analysis," Chapter 7). Whiting (1998) and Siddal and Whiting (1999) emphasized that the 13 taxa analyzed by Carmean and Crespi and by Huelsenbeck clearly represent an inadequate sample for reconstructing relationships among the holometabo-lous insects. The fact that Strepsiptera — and Diptera — always group with other Antliophora (Mecoptera, Siphonaptera) would, as stated by Whiting et al., there-fore suggest relationship, rather than artifact of method. The conclusions of Huelsenbeck represent little more than another attempt in the name of long-branch attraction to justify the use of maximum-likelihood methods and the necessarily arbitrary choice of models of sequence evolution, rather than ac-cepting at face value a parsimonious evaluation of all available data. (See also Siddal, 1998.)

The degree to which the idea of long-branch attraction has been uncritically accepted is exemplified in a sophistic disquisition on the subject by Lyons-Weiler and Hoelzer (1997). Three aspects of that discussion attract particular at-tention.

First, the authors stated that it is *known* that long branches attract and that the accuracy of tree estimation can be greatly improved when the long branches are identified and their harmful influences mitigated. Mitigation, in the sense used by these authors, involves analyzing only a subset of characters, employing alternative characters such as amino acid sequences instead of DNA sequences, and strategically sampling or *excluding* taxa to reduce long branches. These ideas are not new to these authors, however, as they can also be found in the work of Swofford et al. (1996:427): "the only hope of getting the correct tree is by sam-pling few enough characters that we may be lucky enough to obtain more of the character patterns favoring the true tree."

Second, in the view of Lyons-Weiler and Hoelzer, long branches can mislead tree estimation because taxa at the ends of those branches exhibit character states that no longer retain genealogical information. This statement is intro-duced into the paper as if it applied to all data used in phylogenetic analyses, whereas further reading indicates that it refers only to DNA sequence data. To a morphologist, steeped in the concept of character transformation, such a state-ment might be considered pure nonsense. But any worker, irrespective of train-ing, could conclude on the basis of logic that if genealogical information is not retained, the phylogenetic relationships of those "long-branch taxa" could never be determined.

Third, Lyons-Weiler and Hoelzer observed that parsimonious reconstruction of character-state transformations on an evolutionary tree can grossly under-estimate rates of evolutionary change when the assumption of parsimony does not apply. The context of this statement, and additional comments in the article, make it patently clear that the authors believe that applying parsimony methods implies belief in parsimonious evolution, an idea entertained by no serious students of the subject.

The article by Lyons-Weiler and Hoelzer represents an example of the perversion of logic in an attempt to justify suppositions about the properties of methods. Possibly most startling is the idea that fewer data points will somehow corroborate hypotheses of greater information content. Such an approach would not be within the province of science, but only of belief because it would pave the way to discarding any data that run contrary to preconception.

Literature Cited

Archie, J. W. 1989. Homoplasy excess ratios: new indices for measuring levels of homoplasy in phylogenetic systematics and a critique of the consistency index. *Syst. Zool.* 38:253–269.

Camin, J. H., and R. R. Sokal. 1965. A method for deducing branching sequences in phylogeny. *Evolution* 19:311–326.

Carmean, D., and B. Crespi. 1995. Do long branches attract flies? *Nature* 373:666.

Carpenter, J. M. 1988. Choosing among multiple equally most parsimonious cladograms. *Cladistics* 4:291–296.

Carpenter, J. M. 1990. On genetic distances and social wasps. *Syst. Zool.* 39:391–397.

Day, W. H. E., D. S. Johnson, and D. Sankoff. 1986. The computational complexity of inferring rooted phylogenies by parsimony. *Math. Biosci.* 81:33–42.

Edwards, A. W. F., and L. L. Cavalli-Sforza. 1964. Reconstruction of evolutionary trees. pp. 67–76. *In:* Heywood, V. H., and J. McNeill (eds.), *Phenetic and Phylogenetic Classification.* Systematics Association Publication No. 6., London.

Estabrook, G. F., J. G. Strauch, Jr., and K. L. Fiala. 1977. An application of compatability analysis to the Blackith's data on orthopteroid insects. *Syst. Zool.* 26:269–276.

Farris, J. S. 1969. A successive approximations approach to character weighting. *Syst. Zool.* 18:374–385.

Farris, J. S. 1970. Methods for computing Wagner trees. *Syst. Zool.* 19:83–92.

Farris, J. S. 1977. Phylogenetic analysis under Dollo's Law. *Syst. Zool.* 26:77–88.

Farris, J. S. 1979. The information content of the phylogenetic system. *Syst. Zool.* 28:483–519.

Farris, J. S. 1982. Simplicity and informativeness in systematics and phylogeny. *Syst. Zool.* 31:413–444.

Farris, J. S. 1988. *Hennig*86. Documentation.

Farris, J. S. 1989. The retention index and rescaled consistency index. *Cladistics* 5:417–419.

Farris, J. S., and A. G. Kluge. 1979. A botanical clique. *Syst. Zool.* 28:400–411.

Farris, J. S., V. A. Albert, M. Kallersjo, D. Lipscomb, and A. G. Kluge. 1996. Parsimony jackknifing outperforms neighbor-joining. *Cladistics* 12:99–124.

Farris, J. A., et al. 1995. Explanation. *Cladistics* 11:211–218.

Felsenstein, J. 1973. Maximum likelihood and minimum-steps methods for estimating evolutionary trees from data on discrete characters. *Syst. Zool.* 22:240–249.

Felsenstein, J. 1978. The number of evolutionary trees. *Syst. Zool.* 27:27–33.

Fitch, W. M. 1971. Toward defining the course of evolution: minimum change for a specific tree topology. *Syst. Zool.* 20:406–416.

Gladstein, D. S. 1997. Efficient incremental character optimization. *Cladistics* 13:21–26.

Goloboff, P. A. 1993. Estimating character weights during tree search. *Cladistics* 9:83–91.

Goloboff, P. A. 1994. Character optimization and calculation of tree lengths. *Cladistics* 9:433–436.

Goloboff, P. A. 1996. Methods for faster parsimony analysis. *Cladistics* 12:199–220.

Goloboff, P. A. 1998. *Principios básicos de cladistica.* Sociedad Argentina de Botanica, Buenos Aires. 81 pp.

Guttmann, W. F. 1977. Phylogenetic reconstruction: theory, methodology, and application to chordate evolution. pp. 645–669. *In:* Hecht, M. K., P. C. Goody, and B. M. Hecht (eds.), *Major Patterns of Vertebrate Evolution.* Plenum Press, New York.

Huelsenbeck, J. P. 1997. Is the Felsenstein zone a fly trap? *Syst. Biol.* 46:69–74.

Kitching, I. J., P. L. Forey, C. J. Humphries, and D. M. Williams. 1998. *Cladistics: The Theory and Practice of Parsimony Analysis,* second edition. Oxford University Press, Oxford. 228 pp.

Kluge, A., and J. S. Farris. 1969. Quantitative phyletics and the evolution of anurans. *Syst. Zool.* 18:1–32.

Le Quesne, W. J. 1969. A method of selection of characters in numerical taxonomy. *Syst. Zool.* 18:201–205.

Lipscomb, D. L. 1992. Parsimony, homology, and the analysis of multistate characters. *Cladistics* 8:45–65.

Lyons-Weiler, J., and G. A. Hoelzer. 1997. Escaping from the Felsenstein zone by detecting long branches in phylogenetic data. *Mol. Phyl. Evol.* 8:375–384.

Maddison, D. 1991. The discovery and importance of multiple islands of most parsimonious trees. *Syst. Zool.* 40:315–328.

Mickevich, M. F., and D. Lipscomb. 1991. Parsimony and the choice between different transformations of the same character set. *Cladistics* 7:111–139.

Nelson, G., and N. I. Platnick. 1991. Three-taxon statements: a more precise use of parsimony? *Cladistics* 7:351–366.

Nixon, K. C. 1998. The Parsimony Ratchet, a new method for rapid parsimony analysis and broad sampling of tree islands in large data sets. Abstract of a paper presented at the American Museum of Natural History, May 22, 1998.

Platnick, N. I. 1989. An empirical comparison of microcomputer parsimony programs, II. *Cladistics* 5:145–161.

Saitou, N., and M. Nei. 1987. The neighbor-joining method: a new method for reconstructing phylogenetic trees. *Mol. Biol. Evol.* 4:406–425.

Sankoff, D., and R. Cedergren. 1983. Simultaneous comparison of three or more sequences related by a tree. pp. 253–264. *In:* Sankoff, D., and J. Kruskall (eds.), *Time Warps, String Edits, and Macromolecules: The Theory and Practice of Sequence Comparison.* Addison-Wesley, Reading, Massachusetts.

Sankoff, D., and P. Rousseau. 1975. Locating the vertices of a Steiner tree in an arbitrary space. *Math. Program.* 9:240–246.

Sharkey, M. J. 1989. A hypothesis-independent method of character weighting for cladistic analysis. *Cladistics* 5:63–86.

Siddall, M. E. 1998. Success of parsimony in the four-taxon case: Long-branch repulsion by likelihood in the Farris Zone. *Cladistics* 14:209–220.

Siddall, M. E., and A. G. Kluge. 1997. Probabilism and phylogenetic inference. *Cladistics* 13:313–336.

Siddall, M. E., and M. F. Whiting. 1999. Long-branch abstractions. *Syst. Biol.* In press.

Sober, E. R. 1988. *Reconstructing the Past. Parsimony, Evolution, and Inference.* The MIT Press, Cambridge, Massachusetts. 265 pp.

Soltis, D. E., P. S. Soltis, M. E. Mort, M. W. Chase, V. Savolainen, S. B. Hoot, and C. M.

Morton. 1998. Inferring complex phylogenies using parsimony: an empirical approach using three large DNA data sets for angiosperms. *Syst. Biol.* 47:32–42.

Sorensen, J. T., B. C. Campbell, R. J. Gill, and J. D. Steffan-Campbell. 1995. Non-monophyly of Auchenorrhyncha (Homoptera), based upon 18S rDNA phylogeny: eco-evolutionary and cladistic implications within pre-Heteropterodea Hemiptera (s.l.) and a proposal for new monophyletic suborders. *Pan-Pac. Entomol.* 71:31–60.

Swofford, D. L., and W. P. Maddison. 1987. Reconstructing ancestral character states under Wagner parsimony. *Math. Biosci.* 87:199–229.

Swofford, D. L., G. J. Olsen, P. J. Waddell, and D. M. Hillis. 1996. Phylogenetic Inference. pp. 407–514. *In:* Hillis, D. M., C. Moritz, and B. K. Mable (eds.), *Molecular Systematics.* Second Edition. Sinauer Associates, Sunderland, Massachusetts.

Wenzel, J. W. 1997. When is a phylogenetic test good enough? pp. 31–54. *In:* Grandcolas, P. (ed.), *The Origin of Biodiversity in Insects: Phylogenetic Tests of Evolutionary Scenarios.* Vol. 173. Mem. Mus. Natn. Hist. Nat., Paris.

Whiting, M. F. 1998. Long-branch distraction and the Strepsiptera. *Syst. Biol.* 47:134–138.

Whiting, M. F., and W. C. Wheeler. 1994. Insect homeotic transformation. *Nature* 368:696.

Whiting, M. F., J. C. Carpenter, Q. D. Wheeler, and W. C. Wheeler. 1997. The Strepsiptera problem: phylogeny of the holometabolous insect orders inferred from 18s and 28s ribosomal DNA sequences and morphology. *Syst. Biol.* 46:1–68.

Suggested Readings

Farris, J. S. 1970. Methods for computing Wagner trees. *Syst. Zool.* 19:83–92. [The classic paper on phylogenetic algorithms]

Farris, J. S. 1983. The logical basis for phylogenetic analysis. pp. 1–36. *In:* Platnick, N. I., and V. A. Funk (eds.), *Advances in Cladistics,* Volume 2. Proceedings of the Second Meeting of the Willi Hennig Society. Columbia University Press, New York. [A critical discussion of the logic of phylogenetic inference]

Goloboff, P. A. 1993. Estimating character weights during tree search. *Cladistics* 9:83–91. [A useful discussion of weighting in general, and implied weights in particular]

Swofford, D. L., G. J. Olsen, P. J. Waddell, and D. M. Hillis. 1996. Phylogenetic Inference. pp. 407–514. *In:* Hillis, D. M., C. Moritz, and B. K. Mable (eds.), *Molecular Systematics,* second edition. Sinauer Associates, Sunderland, Massachusetts. [An extended and ecelctic discussion of approaches to phylogeny reconstruction]

7

Evaluating Results

In this chapter we will examine the tools used to evaluate the results of phylogenetic analyses. Some are broadly applied and generally understood as valid. Others have been less widely applied, although conforming to the philosophy of science advocated in Chapter 3. Still others appear to have serious limitations or do not conform to the hypothetico-deductive approach at all.

Multiple Equally Parsimonious Cladograms

Even if the method we use for analyzing our data always produces a single tree, it will be of little use if the tree does not represent an accurate evaluation of the data. We mentioned the "neighbor joining" method in this regard in Chapter 6 (Islands of Trees and Solutions for Very Large Data Sets), which produces a single tree, but under no specified optimality criterion and without regard to the number of trees the data might actually imply. Many phenetic techniques have similar properties. In neither case do the advantages of having an unqualified answer outweigh the fact that the answer tells us little about the attributes of the data themselves. The logic of applying parsimony in cladistic analysis is justified — as explicated in much of the foregoing analysis — because the hypotheses produced under the criterion are as closely related to the data as possible. Although applying parsimony may produce results in the form of one or a few trees, the numbers of trees can be much higher. The following techniques have been proposed for use in further evaluation of results containing multiple trees.

Consensus and Compromise Techniques

The consensus approach was proposed by Adams (1972) as a way of combining information, from rival classifications or rival trees, for the same set of organisms. Subsequently, a number of additional techniques have been described: until recently their results were lumped under the term "consensus tree." All members of this class of techniques search for "sets" of taxa common among trees, differing only in the strictness of their requirement for correspondence of the contents of those sets with the original trees. Nixon and Carpenter (1996) suggested that application of the term "consensus tree" should be restricted to the "strict consensus" technique, which produces an output tree containing only groups found in *all* input trees. They recommended calling the output from all other techniques "compromise trees," because even though those approaches often produce more highly resolved output trees than is the case for strict consensus, those trees do not summarize exactly groupings from all of the input trees. The following definitions and Fig. 7.1 summarize the attributes of available techniques. In the definitions, "component" refers to a node in a cladogram and all of the branches descending from it. The confused history of terminology concerning consensus and compromise approaches was reviewed by Nixon and Carpenter (1996).

Consensus technique (strict; Nelson). In this approach, only the monophyletic groups occurring in all input trees are found in the resultant tree. This method was used by Schuh and Polhemus (1980) and attributed to Nelson (1979) by them; the results were later termed "Nelson trees" by Schuh and Farris (1981) and "strict consensus trees" by Sokal and Rohlf (1981). The phylogenetics software packages HENNIG86 and PEE-WEE/NONA use the command "nelsen" to produce such trees.

Compromise techniques
- *Combinable components (Bremer):* The resultant tree contains only monophyletic groups found in at least one of the input trees, but is compatible (non-conflicting) with all of the input trees (Bremer, 1990).
- *General (Nelson, after Page, 1989):* The resultant tree contains all replicated components and all nonreplicated components combinable with the replicated components and with each other (Nelson, 1979).
- *Adams:* Unstable components and taxa are pulled down to the first node on the cladogram that summarizes the different topologies (Adams, 1972). This result may contain groups that appear in none of the input trees.
- *Majority rule:* The resultant tree contains the monophyletic groups found in over 50 percent of the input trees (Margush and McMorris, 1981). This is the weakest of all possible techniques because groupings incongruent with the compromise tree may exist in the 49.99 percent of the trees not used to form the compromise.

Consensus results have been criticized as not providing the most efficient explanation of the data on which the input cladograms were based (e.g., Miyamoto, 1985). Bremer (1990) argued that they are, nonetheless, *useful for forming classifications* because they summarize information on groups (the recognition

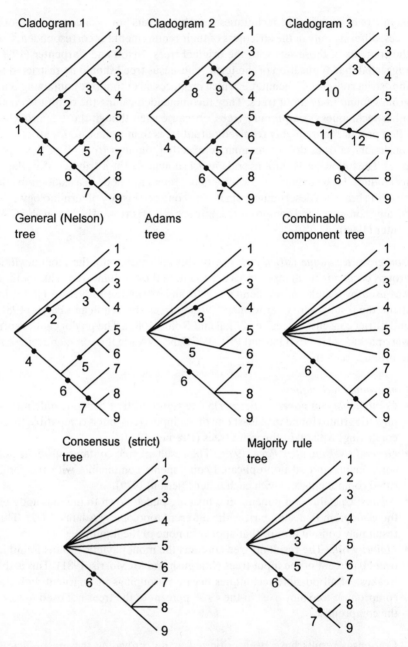

Fig. 7.1. Results of consensus and compromise techniques. *Top:* three input cladograms; *bottom:* consensus and compromise results under five methods (modified from Bremer, 1990). These techniques are discussed in the text.

of taxonomic congruence) for which there is support in the data. This aspect of their utility exists in those not-infrequent situations where multiple cladograms occur and where other parsimony-consistent techniques do not resolve the choice among them because the structure of the data themselves do not permit it.

The true consensus may often not be well resolved, but common information on grouping will nonetheless be faithfully reproduced. Compromise techniques all fail — to a greater or lesser degree — to transmit information on groupings found in all of the input trees. As noted above, the results of Adams' technique may even include groups not found in any of the input trees, and the majority rule approach may have ignored information in direct conflict with the result.

The true consensus contains only those branches that unambiguously support all possible optimizations of the character data, as was pointed out by Nixon and Carpenter (1996). This will not be true of the results of compromise techniques. Therefore, the consensus technique is the obvious preference among all techniques proposed because its results are the only ones that receive unequivocal evidential support.

Successive Approximations Weighting

Carpenter (1988) recommended successive approximations weighting as an approach to selecting among — and thereby reducing — the numbers of equally most parsimonious trees often found when using parsimony algorithms in phylogenetic analysis. He argued that the results would be superior to those of the consensus approach because there would be one or a few trees that described the data optimally, rather than a single tree of suboptimal explanatory power. The approach for implementing and the logic behind use of successive weighting have already been described in Chapter 6.

Many authors have applied successive approximations weighting under the approach advocated by Carpenter and shown that the method often greatly reduces the total numbers of most parsimonious trees. Nonetheless, an apparently confounding result may be one containing none of the most parsimonious trees to which successive weighting was applied. Such a result could be considered ambiguous and disregarded; however, there may be reasons to consider one or more such trees as the most desirable of all trees produced during analysis of the data.

As an example, consider the arguments of Platnick et al. (1991), who, while conducting a phylogenetic analysis of haplogyne spider relationships, found 10 cladograms of equal length, on the basis of 67 characters for 43 taxa. When these results were successively weighted, the authors found six cladograms with stable weights, none of which were in the initial set of 10. Of those six, Platnick et al. deemed one preferable because it was among the most highly resolved and therefore the most informative of the six, and was — on the basis of the un

Sidebar 9
Topology: The Shapes of Cladograms

Practitioners of phenetics have at times argued for the desirability of techniques producing trees with symmetrical branching. But it has been shown that there is no reason to expect symmetrical trees over asymmetrical, or vice versa (Farris, 1976). Thus, there would seem to be no reason to include tree shape as a justification for choice of method. Whether phylogenetic relationships actually take one form more commonly than another is irrelevant for phylogenetic methods, but asymmetry in trees appears to be quite common, nonetheless. Asymmetrical cladograms imply a greater degree of "nesting" of synapomorphic characters, and therefore a greater degree of implied character support for the more distal branch points than would be the case for a symmetrically branching tree with the same number of taxa (see, e.g., Mickevich and Platnick [1989], concerning the information content of classifications).

Tree-like structures depicting phylogenetic relationships have been drawn in many different forms. The myriad of possible cladograms for even a limited number of taxa makes it desirable to draw cladograms to facilitate ease of comparison of branching patterns. One convention — although not universal — is for cladograms to always branch to the right, as shown in Fig. 7.2.

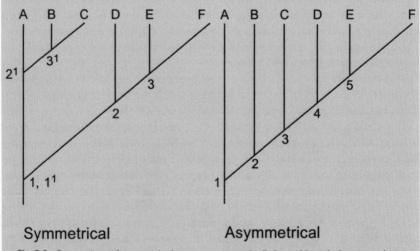

Fig. 7.2. Comparison of symmetrical versus asymmetrically branching cladograms, showing the greater degree of nesting of synapomorphies in the asymmetrical topology on the right.

weighted data — only one step longer than the original 10, whereas others among the six were as much as three steps longer. The choice of that tree was further justified because it was also the one produced under the "fittest tree" approach of Goloboff (Platnick et al., 1996). Whereas nearly all previously published analyses had accepted the most parsimonious tree(s) as the answer with greatest explanatory power, and therefore preferred, Platnick et al. (1991) argued that treating all characters as being of equal weight biased the result in favor of some characters of lesser explanatory power.

Data Decisiveness

It has been argued, erroneously as we saw in Chapter 3, that homoplasy in and of itself limits our ability to conduct phylogenetic analysis under parsimony. A measure of *data decisiveness* was developed by Goloboff (1991) as a way of distinguishing amount of homoplasy from confidence in results. Goloboff showed that decisiveness (or lack thereof) has no direct relationship to homoplasy. Substantially homoplasious data may be relatively decisive, that is, allow for a clear choice among cladograms. This conclusion is concordant with the observation of Farris, that just because parsimony attempts to minimize character change, parsimony can nonetheless be applied when character change is not minimal. What data decisiveness does show is that there is greater reason to prefer a result in which one tree with relatively high homoplasy is generated as opposed to a result comprising thousands of equally parsimonious and nearly most parsimonious trees, even though those trees have a relatively high consistency index. We may therefore conclude that a large amount of homoplasy does not, ipso facto, imply minimal confidence in results.

Separate versus Combined Analysis: Total Evidence versus Consensus

Some authors have argued that we should treat morphological versus molecular, larval versus adult, and morphological versus behavioral data as separate sources of evidence for a given set of taxa. Assessment of the degree to which these data types describe the same set of relationships for a given group of taxa would seem to be a central problem for systematics. Examination of arguments on the proper methods for addressing this issue allows for the identification of essentially two opposing approaches — indeed philosophies:

- *Total evidence.* Proponents of this viewpoint argue that all data for a set of taxa should be pooled and analyzed simultaneously. The rationale is that only through such an approach can global agreement among data for the set of taxa be properly assessed. The term "total evidence" was applied by Kluge (1989) and Kluge and Wolf (1993) in conjunction with their original arguments for the approach. "Combined analysis" and "simultaneous analysis" have also been recommended as possibly more appropriate labels (Nixon and Carpenter, 1996).

- *Consensus.* The alternative viewpoint — sometimes called "taxonomic congruence" — is that different data sets for the same set of taxa should be analyzed independently, and commonality of conclusion should be judged via the application of consensus. The primary arguments justifying this approach are that (1) "independent" data should be subject to independent analyses; (2) discrete character data and distance data cannot be analyzed simultaneously in their native form and therefore demand the use of consensus; and (3) DNA sequence data may "swamp" morphological data if the two are analyzed together because of the much greater number of data points involved in the former.

In order to clarify the history of usage, it should be noted that the term *taxonomic congruence* was used by Mickevich (1978) in reference to the degree to which classifications based on different data sets imply the same conclusions, an idea originally proposed in the literature on phenetics. The work of Mickevich and others (e.g., Schuh and Polhemus, 1980; see also Schuh and Farris, 1981) used consensus techniques in attempts to refute claims that phenetic approaches produced more informative and stable results than cladistic techniques. In the work of Mickevich (1978), the data were of ostensibly different types (e.g., morphological and allozyme); in the case of Schuh and Polhemus, subsets of the total data set were selected at random. Thus, the more recent usage of "taxonomic congruence" as a synonym of "consensus" confounds approaches with very different justifications. The differences between the three approaches are clarified in Fig. 7.3. It should be obvious from what has been said so far that "total evidence" and "consensus" are approaches to judging data. "Taxonomic congruence," in the sense of Mickevich, refers to an approach for judging the efficacy of methods. Indeed, for Mickevich (1978), the question was not whether different data sets would tell the same story — she assumed it — but rather how to judge the ability of different approaches to produce informative and stable classifications based on those data.

Having clarified the terminology and outlined some basic tenets, how then might we judge the validity of the arguments for and against the total evidence and consensus approaches?

Evaluating the Total Evidence Approach

The justification for total evidence is essentially the argument for parsimony as an integral part of method in science. There also appear to be sound empirical reasons for combining data, such as the observation that combined analysis of data sets with substantial homoplasy may produce results showing greater congruence than separate analyses of the same data sets.

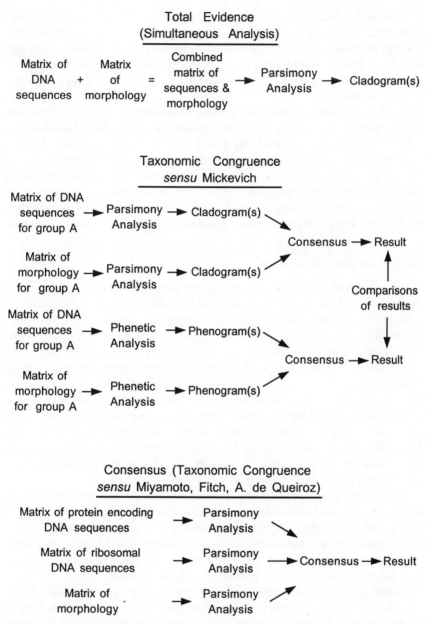

Fig. 7.3. Comparison of three analytic approaches to cladistic analysis: *total evidence* (simultaneous analysis) as originally proposed by Kluge and Wolf (1993); *taxonomic congruence* as used by Mickevich (1978) to judge the ability of methods to produce congruent classifications from different data sets for the same taxa; and *consensus* as used by Miyamoto, Fitch, and de Queiroz, Donoghue, and Kim to evaluate the conformity of multiple data sets to a preconceived "true" answer.

Arguments against total evidence are of two types. First, cladograms derived from certain data sets may not reflect the *true* phylogeny for the group, a "fact" that will be obscured by the total evidence approach. Our discussions of the philosophy of science will force us to reject this argument because we can never know the "true" tree; if we did, we would have no need for the results of phylogenetic analysis.

Second, it has been argued that because all data types cannot be combined, as, for example, distance data and discrete character data, some data may have to be excluded from a total evidence analysis (de Queiroz, Donoghue, and Kim, 1995; Miyamoto and Fitch, 1995). Combining such data would, per force, require the use of consensus techniques. The exceptional cases necessitating the use of consensus certainly offer no necessary justification for analyzing all data subsets separately and then seeking the final answer via consensus. In those individual cases where analysis via consensus is necessary, that method should be applied.

Evaluating the Consensus Approach

The primary argument for consensus seems to be that the aim of phylogenetic studies is to find the "true" phylogeny and that not all data sets are equally capable of revealing that result (de Queiroz, Donoghue, and Kim, 1995; Miyamoto and Fitch, 1995). On this view, then, if all data are combined in a single matrix, the correspondence — or lack thereof — of subsets of the data to the "true" phylogeny will be obscured. Knowledge of the "true" phylogeny, according to these authors, may be derived from simulations, studies of well-supported phylogenies of natural groups, and known phylogenies of laboratory bred and domesticated groups (Miyamoto and Fitch, 1995). The authors seem unwilling to admit that the "true" phylogeny of the millions of extinct and living species can never be determined through these means. Genealogical relationships in those groups will be understood through the application of cladistic methods, the efficacy of which must be argued on other grounds.

A second argument in favor of consensus is that *independence* is more likely for characters from different data sets than for characters from the same data set and that the importance of independence among data sets justifies using the consensus approach. Yet, the independence of characters will be no greater using consensus than would be the case in a combined data set; thus, the argument does not in and of itself offer a justification for the method.

A third argument, incorporating ideas promoted by Felsenstein and Sober (see Chapter 3), suggests that parsimony may provide misleading results if there is a high degree of homoplasy. In this view, if some data sets (subsets) are highly homoplasious, they should be analyzed separately because, when combined in a simultaneous analysis, the result could lead to an erroneous conclusion. We have argued against this logic at every turn, beginning in Chapter 3. As pointed out above, the argument has no methodological or empirical justification. Indeed, the work of Wheeler, Cartwright, and Hayashi (1993) and Soltis et al.

(1998) indicates that whereas individual molecular data sets for a given group of taxa may be highly homoplasious and produce what might be viewed as meaningless results, combining data from different DNA regions for a group of taxa can produce a much stronger phylogenetic signal.

Fourth, studies combining disparate data types have failed to support the belief that sequence data will overwhelm or "swamp" morphological data and therefore influence the result disproportionately (e.g., Chippindale and Wiens, 1994). If we assumed that all sites in a DNA sequence varied in an informative way across the set of taxa being analyzed, then sequence data might overwhelm morphology in any given analysis. All empirical evidence shows, however, that the number of informative sites is probably on the order of 20 percent or less for most data sets and that among those sites there is still substantial homoplasy. Many combined morphology–DNA sequence data sets contain a proportion of morphological characters roughly equal to 20 percent of the total number of nucleotides being analyzed. Furthermore, the total evidence (simultaneous analysis) approach, in principle, argues that data set size should be irrelevant and that character congruence should be the final arbiter for judging the phylogenetic value of data.

Arguing against consensus, some authors have emphasized that the approach does not produce results that optimally describe the data because the consensus result is usually less well resolved than any one or all of the input phylogenies. Thus, information on characters is lost (Miyamoto, 1985), even though information on shared groupings is not. This is essentially the justification of total evidence given above.

In summary, the only strong argument for taxonomic congruence would seem to be in the sense that the term was used by Mickevich (1978), as a test of the efficacy of methods. All other arguments for consensus as an analytic technique appear to be based on preconceptions about the nature of the data and of the expected results. Indeed, a substantial body of empirical data is now accumulating with the strong suggestion that simultaneous analysis of molecular + molecular data sets or molecular + morphological data sets produces results with higher consistency and retention indices, the strongest indicators of hierarchic structure within the data. This conclusion does not, however, negate the value of consensus for revealing groupings for which unequivocal character support exists in a data set and for the recognition of those groups in formal classifications.

Measures of Support for Results

Approaches for evaluating confidence in phylogenetic results have been devised on the assumption that we might wish insight beyond the application of parsimony alone. Some measures purporting to provide such insight are statistically based, others are not.

Sidebar 10
Gene Trees versus Species Trees: Wherein Lies the Holy Grail?

The distinction between "species trees" and "gene trees" was first introduced by Morris Goodman (viz., Goodman et al., 1979), but was implicit
in Fitch's (1970) writings on paralogous genes. A *species tree* in Goodman's
view was one indicating a "known" scheme of phylogenetic relationships.
A *gene tree,* by comparison, described a scheme of relationships — based
on genetic components such as amino acid or DNA sequences — that did
not conform to the species tree. In the work of Goodman et al. (1979),
the "true" phylogeny — species tree — was proposed on the basis of non-
sequence data; what data Goodman et al. actually used to derive that
scheme of relationships is unclear in their paper. More recently, Maddison
(1997) discussed this issue, *assuming* a "true phylogeny," but without indicating how it might be recognized as distinct from any possible scheme
of phylogenetic relationships.

The original idea of Goodman was extended, by way of explanation,
when Tajima (1983) noted that — for certain genes — within-group variation can be greater than between-group variation, even in mitochondrial
genes where recombination is negligible. Tajima concluded that it is therefore possible to infer incorrect phylogenetic conclusions when certain
nucleotide sequence data are analyzed by themselves; these results he referred to as *gene trees.* Further examples of processes that produce gene
trees not corresponding to species trees are, according to Doyle (1992):
(1) horizontal gene transfer, (2) introgression, (3) ancestral polymorphism,
and (4) paralogous gene families as originally envisioned by Fitch and
Goodman et al.

The validity of the gene tree–species tree argument was questioned by
Brower, DeSalle, and Vogler (1996:434) when they noted that the "attention paid to them [gene trees] in the literature may reflect more the unease
they inspire in our phylogenetic paradigm than their prevalence in nature."
These authors further stated that a significant number of examples citing
the confounding nature of "gene tree" data are derived from analyses
whose methods invalidate the conclusions in the first place, or were based
on unrealistic evolutionary models, or for whose use there was no justification. We might add that the question of whether population-level variation is amenable to phylogenetic analysis had already been addressed
by Hennig, who concluded that variation at this level is nonhierarchic, or
tokogenetic. Many examples of introgression would appear to fit in this
category.

The gene tree–species tree dichotomy is strongly associated with the
idea that phylogenetic analysis aims to discover the *true* tree of life. As discussed in Chapter 3, such an approach would seem to be on philosophi-

cally and scientifically shaky ground. As noted by Brower et al. (1996:442), "If we believe there is a hierarchical pattern, then we should expect the data to reflect that pattern alone . . . The best estimate of hierarchical relationships is still derived from parsimonious interpretation of all the data." Alternatively stated, the strongest test of valid comparison is the recovery of a hierarchic pattern.

The discovery of that pattern, whether on the basis of gross morphological data, behavioral data, or molecular-level data, ultimately rests on theories of character homology and transformation. If we assume that duplicated portions of the genome — paralogs — cannot on their face be distinguished from one another and will therefore be associated erroneously, we will have compared nonhomologs. Obviously, if we knew the comparisons to be made were erroneous, we would not make them in the first place (see discussions in de Pinna, 1991; Vrana and Wheeler, 1992). The strongest test of homology theories is in the discovery of congruence. Therefore, incongruent schemes might suggest comparisons of nonhomologous sequences. But, we find that the gene tree–species tree argument has frequently been used to explain perceived incongruence, *a priori.*

Bremer Support

As a way of determining support for branches as implied by the original data, Bremer (1988, 1994; see also application in Davis, 1995) proposed a measure based on the number of extra steps required to lose a branch in the consensus of nearly most parsimonious trees. The idea can be stated alternatively as the difference in length between the most parsimonious trees and the trees one or more steps longer in which the clade is not resolved. The results, often referred to as "Bremer Support values," are usually reported as an integer for each branch in the consensus of trees one step longer than most parsimonious, two steps longer, and so on (Fig. 7.4). The higher the Bremer Support number, presumably the stronger the support for a given clade. The meaning of the number, however, is not entirely clear because of the complexity of interactions among character data. Also, because it is difficult to find all of the trees that are some given number of steps longer than most parsimonious, individual phylogenetics programs may report different numbers of nearly most parsimonious trees, which will inevitably influence the numbers reported as Bremer Support values. The Bremer Support number will not necessarily be the same as the number of steps longer than most parsimonious required to loose a clade in the consensus.

Jackknifing

The use of the jackknife was proposed by Lanyon (1985; see additional discussion in Siddall, 1995) as a method assessing inconsistencies in distance data. The

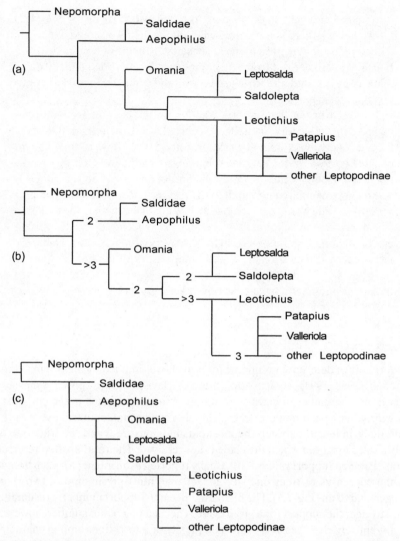

Fig. 7.4. Bremer Support values for the Leptopodomorpha data set of Schuh and Polhemus (1980) as generated by Goloboff's program NONA. *a.* Most parsimonious tree. *b.* Tree with Bremer Support values for trees three steps longer than most parsimonious. *c.* Consensus of nine trees used to calculate Bremer Support values given in *b.*

jackknife, as implemented by Lanyon, produces "pseudoreplicates" in which one taxon (or more) is removed from the data matrix for each sample, that taxon being replaced and another withdrawn for the next sample, until all such combinations of samples have been analyzed. Jackknife results are reported as a proportion of the cladograms in which a clade occurs (Fig. 7.5). Of all data sub-sampling and randomization methods so far proposed, the jackknife may offer

tread
(0 (14 ((13 (11   12))(3 5 6 10 (7 (8 9)) (4 (1 2)))))) ;
proc/ ;

CONSENSUS TREE: 8 NODES (CUTOFF VALUE: 50)
Frequencey index: 0.528

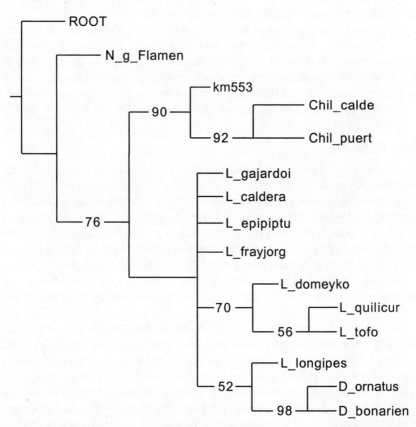

Fig. 7.5. Jackknife results for a small data set as produced through the use of Goloboff's program NONA and a jackknife utility program, based on 50 replications. The consensus contains those clades for which there was jackknife support in more than 50 percent of the trees (the cutoff value). The numbers on the nodes represent the percentage of trees in which the nodes occurred.

the greatest appeal because it lacks the flaws of the others. Namely, it does not attribute support to data where none exists, it is not affected by the presence of autapomorphies, and it does not produce variable results for different cladistically perfect clades just because they possess varying numbers of synapomorphies. (See also description of Parsimony Jackknifing under Islands of Trees in Chapter 6.)

Bootstrapping

The bootstrap, as applied to phylogenetic analysis by Felsenstein (1985), functions by "resampling with replacement." A bootstrap begins by randomly sampling characters from a matrix and maintaining the original size of the matrix by replicating characters in the sample via weighting of the sampled characters; cladograms are then computed. Multiple resamplings are conducted. The results are reported as a proportion of cladograms in which a group occurs, this measure being used to establish a confidence interval for the results.

Although the bootstrap has been widely applied in the phylogenetic literature, its drawbacks have just as frequently been overlooked. The most obvious limitations appear to be that the results are negatively affected by the presence of unique characters (autapomorphies) which serve no function in forming groupings (Carpenter, 1996) and that clades can vary in their respective bootstrap values, even in the absence of homoplasy (Siddall, 1995:47).

Randomization Tests

Yet another proposed approach to assessing confidence in phylogenetic results involves randomizing data. The underlying idea is that apparently informative data might have a structure that could be matched by chance alone. This approach, independently proposed by Archie (1989) and Faith and Cranston (1991), reshuffles (permutes) states of a character randomly among taxa, with each character being treated independently during the course of the permutation process. Multiple permutations of the data are performed, cladograms computed, and comparisons made between cladograms derived from randomized and original data in order to establish confidence limits, usually at the 95 percent level.

Two types of comparisons have been made: those of lengths of the resultant trees (PTP, permutation tail probability) and those dealing with monophyletic groups (T-PTP, topology dependent permutation tail probability). Faith and Cranston (1991) claimed that permutation tests provided an "absolute criterion" by which phylogenetic hypotheses can be judged. As pointed out by Carpenter (1992), the methods provide no test of the hypotheses themselves and make the apparently unwarranted assumption that the number and frequency of character states remain constant. Faith (1992) subsequently offered a justification of his method in a hypothetico-deductive context, suggesting that "what is passed off as 'corroboration' [in cladistics] does not extend beyond identification of the mpt [most parsimonious tree], and cannot be Popperian corroboration at all." This argument was ridiculed by Farris (1995) as being a misinterpretation of the arguments of Popper and a conflation of content of hypotheses and corroboration of hypotheses. Carpenter, Goloboff, and Farris (1998) concluded that neither PTP nor T-PTP are useful in phylogenetic analysis because they can attribute significance to data that support no resolved grouping.

Literature Cited

Adams, E. N., III. 1972. Consensus techniques and the comparison of taxonomic trees. *Syst. Zool.* 21:390–397.

Archie, J. W. 1989. A randomization test for phylogenetic information is systematic data. *Syst. Zool.* 38:239–252.

Bremer, K. 1988. The limits of amino acid sequence data in angiosperm phylogenetic reconstruction. *Evolution* 42:795–803.

Bremer, K. 1990. Combinable component consensus. *Cladistics* 6:369–372.

Bremer, K. 1994. Branch support and tree stability. *Cladistics* 10:295–304.

Brower, A. V. Z., R. DeSalle, and A. Vogler. 1996. Gene trees, species trees, and systematics: a cladistic perspective. *Ann. Rev. Ecol. Syst.* 27:423–450.

Carpenter, J. M. 1988. Choosing among multiple equally most parsimonious cladograms. *Cladistics* 4:291–296.

Carpenter, J. M. 1992. Random cladistics. *Cladistics* 8:147–153.

Carpenter, J. M. 1996. Uninformative bootstrapping. *Cladistics* 12:177–181.

Carpenter, J. M., P. A. Goloboff, and J. S. Farris. 1998. PTP is meaningless, T-PTP is contradictory: a reply to Trueman. *Cladistics* 14:105–116.

Chippindale, P. T., and J. J. Wiens. 1994. Weighting, partitioning, and combining characters in phylogenetic analysis. *Syst. Biol.* 43:278–287.

Davis, J. I. 1995. A phylogenetic structure for the monocotyledons, as inferred from chloroplast DNA restriction site variation, and a comparison of measures of clade support. *Syst. Bot.* 20:503–527.

de Pinna, M. C. C. 1991. Concepts and tests of homology in the cladistic paradigm. *Cladistics* 7:367–394.

de Queiroz, A., M. J. Donoghue, and J. Kim. 1995. Separate versus combined analysis of phylogenetic evidence. *Ann. Rev. Ecol. Syst.* 26:657–681.

Doyle, J. J. 1992. Gene trees and species trees: molecular systematics as one-character taxonomy. *Syst. Bot.* 17:144–163.

Faith, D. P. 1992. On corroboration: a reply to Carpenter. *Cladistics* 8:265–273.

Faith, D. P., and P. S. Cranston. 1991. Could a cladogram this short have arisen by chance alone?: On a permutation test for cladistic structure. *Cladistics* 7:1–28.

Farris, J. S. 1976. Expected asymmetry of phylogenetic trees. *Syst. Zool.* 25:196–198.

Farris, J. S. 1995. Conjectures and refutations. *Cladistics* 11:105–118.

Felsenstein, J. 1985. Confidence limits on phylogenies: an approach using the bootstrap. *Evolution* 39:783–791.

Fitch, W. M. 1970. Distinguishing homologous from analogous proteins. *Syst. Zool.* 19:99–113.

Goloboff, P. A. 1991. Homoplasy and the choice among cladograms. *Cladistics* 7:215–232.

Goodman, M., J. Czelusniak, G. W. Moore, A. E. Romero-Herrera, and G. Matsuda. 1979. Fitting the gene lineage into its species lineage: a parsimony strategy illustrated by cladograms constructed from globin sequences. *Syst. Zool.* 28:132–163.

Kluge, A. G. 1989. A concern for evidence and a phylogenetic hypothesis of relationships among *Epicrates* (Boidae, Serpentes). *Syst. Zool.* 38:7–25.

Kluge, A. G., and A. J. Wolf. 1993. Cladistics: What's in a word? *Cladistics* 9:183–199.

Lanyon, S. M. 1985. Detecting internal inconsistencies in distance data. *Syst. Zool.* 34:397–403.

Maddison, W. P. 1997. Gene trees and species trees. *Syst. Biol.* 46:523–536.

Margush, T., and F. R. McMorris. 1981. Consensus n-trees. *Bull. Math. Biol.* 43:239–244.

Mickevich, M. F. 1978. Taxonomic congruence. 1978. *Syst. Zool.* 27:143–158.

Mickevich, M. F., and N. I. Platnick. 1989. On the information content of classifications. *Cladistics* 5:33–47.

Miyamoto, M. M. 1985. Consensus cladograms and general classifications. *Cladistics* 1:186–189.

Miyamoto, M. M., and W. M. Fitch. 1995. Testing species phylogenies and phylogenetic methods with congruence. *Syst. Biol.* 44:64–76.

Nelson, G. 1979. Cladistic analysis and synthesis: principles and definitions, with a historical note on Adanson's *Familles des Plantes (1763–1764)*. *Syst. Zool.* 28:1–21.

Nixon, K. C., and J. M. Carpenter. 1996. On simultaneous analysis. *Cladistics* 12:221–241.

Page, R. D. M. 1989. Comments on component compatibility in historical biogeography. *Cladistics* 5:167–182.

Platnick, N. I., J. A. Coddington, R. R. Forster, and C. E. Griswold. 1991. Spinneret morphology and the phylogeny of haplogyne spiders (Araneae, Araneomorphae). *Amer. Mus. Novitates* 3016:73 pp.

Platnick, N. I., C. J. Humphries, G. Nelson, and D. M. Williams. 1996. Is Farris optimization perfect? Three-taxon statements and multiple branching. *Cladistics* 243–352.

Schuh, R. T., and J. S. Farris. 1981. Methods for investigating taxonomic congruence and their application to the Leptopodomorpha. *Syst. Zool.* 30:331–351.

Schuh, R. T., and J. T. Polhemus. 1980. Analysis of taxonomic congruence among morphological, ecological, and biogeographic data sets for the Leptopodomorpha (Hemiptera). *Syst. Zool.* 29:1–26.

Siddall, M. E. 1995. Another monophyly index: revisiting the jackknife. Cladistics 11:33–56.

Sokal, R. R., and F. J. Rohlf. 1981. Taxonomic congruence in the Leptopodomorpha re-examined. *Syst. Zool.* 30:309–325.

Soltis, D. E., P. S. Soltis, M. E. Mort, M. W. Chase, V. Savolainen, S. B. Hoot, and C. M. Morton. 1998. Inferring complex phylogenies using parsimony: an empirical approach using three large DNA data sets for angiosperms. *Syst. Biol.* 47:32–42.

Tajima, F. 1983. Evolutionary relationships of DNA sequences in finite populations. *Genetics* 105:437–460.

Vrana, P., and W. Wheeler. 1992. Individual organisms as terminal entities: laying the species problem to rest. *Cladistics* 8:67–72.

Wheeler, W. C., P. Cartwright, and C. Y. Hayashi. 1993. Arthropod phylogeny: a combined approach. *Cladistics* 9:1–39.

Suggested Readings

Carpenter, J. M. 1988. Choosing among multiple equally most parsimonious cladograms. *Cladistics* 4:291–296. [Discussion of the application of successive approximations weighting]

Goloboff, P. A. 1991. Homoplasy and the choice among cladograms. *Cladistics* 7:215–232. [Discussion of data decisiveness]

Nixon, K. C., and J. M. Carpenter. 1996. On simultaneous analysis. *Cladistics* 12:221–241. [Discussion of the total evidence approach]

APPLICATION OF CLADISTIC RESULTS

8

Formal Classifications and Systematic Databases

In this chapter we will explore two areas that represent the most practical results of phylogenetic analysis — formal classifications and systematic databases.

Formal Classifications

Arguments for Cladistic Classifications

Chapters 4 through 7 of this book dealt with arguments concerning how we bring evidence to bear on the process of phylogeny reconstruction. Hennig argued that there should be a direct correspondence (isomorphy) between phylogenies and formal classifications. The necessity for this correspondence was assessed by Farris (1979) on the basis of efficiency of diagnoses. Farris emphasized that a salient feature of a hierarchy is the impossibility of every group being most informative for every feature considered because some groups are subsets of other groups. Farris demonstrated that in this regard most parsimonious trees (1) allow all data to be summarized in the most succinct diagnoses; (2) minimize exceptions to such diagnoses; and (3) permit the greatest number of predictions between adjacent taxa. Such conclusions correspond closely to Hennig's ideas concerning the role of classifications as the general reference system for biology.

The following examples will serve to illustrate the points made by Hennig and Farris. Consider the traditional classification of the Insecta:

Insecta
  Entognatha
  Ectognatha
    Apterygota
      Machilida (Archaeognatha)
      Thysanura (Zygentoma)

Pterygota
  Odonata
  Ephemeroptera
  Neoptera

The Machilida and Thysanura were long grouped together because they lack wings but nonetheless share many features not found in the more primitive entognathous insects. However, whereas the mandibles in Machilida have a single point of articulation, the Thysanura share with all winged insects the dicondylar structure of the mandibles (among other features). This fact is recognized in the following classification:

Insecta
  Entognatha
  Ectognatha
    Machilida (Archaeognatha)
    Dicondylia
      Thysanura (Zygentoma)
      Pterygota
        Odonata
        Ephemeroptera
        Neoptera

In this scheme the Thysanura are not grouped with the Machilida, but rather are the sister group of all other insects on the basis of mandibular articulation and other distinctive features. Although the first classification would seem to convey information on similarity more effectively, that conclusion is an illusion. Grouping on the absence of distinctive characters allows for any grouping. For example, the absence of insect wings as a group-defining character could group Machilida and Thysanura with any other organism except winged insects. Such groupings would require that all characteristics of all groupings be completely enumerated as a prerequisite to recognizing those groups, an obviously inefficient system. Grouping on the basis of synapomorphy (parsimony) produces the most efficient hierarchic representation of the data because there is minimal repetition of attributes.

Arguments Against Cladistic Classifications

Even though we have seen that parsimony produces the most efficient representation of data on which classifications are based, we might nonetheless wish to examine arguments that have been brought forward against cladograms as the basis of biological classifications and against the application of parsimony more generally.

The most frequently heard objections to cladistic classifications were those of the "classical evolutionary taxonomists" (e.g., Mayr, 1974). They argued that

1. such an approach would require reranking of many familiar taxa;
2. the numbers of ranks would be so great as to be unworkable; and
3. under Hennig's *monophyly* concept, such schemes would require discarding information on similarity, an important aspect of evolution.

The pheneticists argued that

1. information on similarity would be discarded;
2. the cladistic approach would not produce true classifications because it is not based on an ultrametric (see Sidebar 8 for further discussion);
3. the number of levels in the classification would be too great; and
4. the cladistic classifications were generally not symmetrical (balanced) and therefore defective.

Let us examine these complaints for possible substance.

Objections to assigning new ranks to familiar taxa may have their deepest roots in the traditions of taxonomy. Yet, if rank is to connote relationship, objecting to changes in ranks that incorrectly portray relationships would seem to be scientifically counterproductive. This point was made by Gaffney (1979), who concluded that a perfectly stable classification could only represent the perpetuation of ignorance.

In point of fact, *reranking* has largely disappeared as an area of contention because even the most ardent supporters of Hennig's original mandate concerning ranking realized that producing a single unified hierarchy for all of life would involve so many ranks as to be unfeasible. Hennig (1969) himself apparently recognized the difficulty of creating one very large classification with a consistent set of naming conventions and consequently used a numbering system in his classification of the Insecta, a portion of which is shown in Fig. 8.1.

The idea that a classification must be ultrametric could be dismissed as purely definitional, but other arguments are available. Farris (1979) showed that it is impossible for a clustering-level distance (ultrametric) to fit distance data better than the best-fitting path-length distance (see also Sidebar 9). Thus, the most parsimonious path-length interpretation will always produce the most efficient hierarchic description of the data. Just because phenetic methods are based on clustering levels (ultrametrics) does not ipso facto mean that such methods always portray information most efficiently in a hierarchic system. In fact, the conclusion is false.

The two remaining arguments of the pheneticists have an empirical relationship to one another. Whether a classification is phenetic or cladistic in origin, given the same topological relationships of the taxa, the numbers of levels will be the same, a point made by Farris (1979) and one widely appreciated outside phenetic circles. However, if the symmetry of the classification is changed, then the numbers of levels may change. Whether one approach to classification produces results that are more consistently symmetrical than another would seem to

    1. Entognatha
      1.1.   Diplura
      1.2.   Ellipura
      1.2.1.   Protura
      1.2.2.   Collembola
    2. Ectognatha
      2.1.   Archaeognatha (microcoryphia)
      2.2.   Dicondylia
      2.2.1.   Zygentoma
      2.2.2.   Pterygota
      2.2.2.1.   Palaeoptera
      2.2.2.1..1.   Ephemeroptera
      2.2.2.1..2.   Odonata
      2.2.2.2.   Neoptera
      2.2.2.2..1.   Plecoptera
      2.2.2.2..2.   Paurometabola
      2.2.2.2..2.1.   Embioptera
      2.2.2.2..2.2.   Orthopteromorpha
      2.2.2.2..2.2..1.   Blattopteriformia
      2.2.2.2..2.2..1.1.   Notoptera (Grylloblattodea)
      2.2.2.2..2.2..1.2.   Dermaptera
      2.2.2.2..2.2..1.3.   Blattopteroidea
      2.2.2.2..2.2..1.3.1.   Mantodea
      2.2.2.2..2.2..1.3.2.   Blattodea
      2.2.2.2..2.2..2.   Orthopteroidea
      2.2.2.2..2.2..2.1.   Ensifera
      2.2.2.2..2.2..2.2.   Caelifera
      2.2.2.2..2.2..2.3.   Phasmatodea
      2.2.2.2..3.   Paraneoptera
      2.2.2.2..3.1.   Zoraptera
      2.2.2.2..3.2.   Acercaria
      2.2.2.2..3.2..1.   Psocodea
      2.2.2.2..3.2..2.   Condylognatha
      2.2.2.2..3.2..2.1.   Thysanoptera
      2.2.2.2..3.2..2.2.   Hemiptera
      2.2.2.2..3.2..2.2.1.   Heteropteroidea
      2.2.2.2..3.2..2.2.1.1.   Coleorrhyncha
      2.2.2.2..3.2..2.2.1.2.   Heteroptera
      2.2.2.2..3.2..2.2.2.   Sternorrhyncha
      2.2.2.2..3.2..2.2.2.1.   Aphidomorpha
      2.2.2.2..3.2..2.2.2.1.2.   Aphidina
      2.2.2.2..3.2..2.2.2.1.2.   Coccina
      2.2.2.2..3.2..2.2.2.2.   Psyllomorpha
      2.2.2.2..3.2..2.2.2.2.1.   Aleyrodina
      2.2.2.2..3.2..2.2.2.2.2.   Psyllina
      2.2.2.2..3.2..2.2.3.   Auchenorrhyncha
      2.2.2.2..3.2..2.2.3.1.   Fulgoriformes
      2.2.2.2..3.2..2.2.3.2.   Cicadiformes
      2.2.2.2.4.   Holometabola

Fig. 8.1. Sample of the numbered hierarchic classification of the Insecta (excluding details of the Holometabola) from *Die Stammesgeschichte der Insekten* (Hennig, 1969).

be irrelevant for judging a method (see Sidebar 9), unless the *goal* of the method was none other than to minimize the number of hierarchic levels. Such a goal has not been claimed *a priori* for any approach to classification in modern times, and its use as a criticism of cladistic results can hardly be taken seriously because there is no empirical evidence that would justify such an approach.

Observation reveals, nonetheless, that the results of cladistic analyses often do show pronounced asymmetries — what are frequently referred to as *pectinate* or *comblike* cladograms. As we have discussed throughout this book, cladistic classifications, pectinate or otherwise, result from applying methods designed to best describe all available data. Thus, the topological form of the hierarchy in no way reflects on the empirical content of the classification, and cladistic methods do not favor one topological form over another.

If there is a general conclusion to be drawn from 40 years of methodological discussion about biological classification, it is the erroneous nature of the assumption that grouping by overall similarity will provide a more efficient description of information on genealogical relationships than does grouping by synapomorphy. The issue of whether classifications containing only monophyletic groups discard information on similarity, as claimed by both pheneticists and evolutionary taxonomists, would seem to have been unequivocally resolved.

Thus, a recent exegesis by Knox (1998) on the use of hierarchies in systematics cannot help but surprise the reader in its claim that because evolution is dualistic, a hierarchy containing only monophyletic groups cannot therefore represent information on genealogy (phylogenesis) *and* modification (anagenesis). Knox concluded that the systematic research program of the future will involve "setting standards of evidence needed to recognize paraphyletic groups" and "the development of models for analyzing patterns of modification that will complement our current methods of analyzing patterns of descent." Knox's conclusions contain ideas remarkably similar to those of the "evolutionary taxonomist" Ashlock (1979), who advocated paraphyletic groupings in formal classifications, and to those of the pheneticists. Selectively, Knox did not mention the failures of phenetics to achieve its stated objectives, namely, to produce informative, predictive, and natural classifications. Neither did he admit that all of those goals have been realized through the application of cladistic methods. In spite of Knox's assertions to the contrary, cladistic classifications do allow for the effective retrieval of information on genealogy *and* divergence.

By themselves, classifications contain hypotheses about group relationships, which allow us to make predictions concerning the distribution of features unique to the contained groupings (Platnick, 1978). But as has been repeatedly emphasized, the hierarchic representation of taxon relationships does not by itself transmit the information on which the classification is based. That information is in the diagnoses. It is those diagnoses, which are derived through the application of parsimony, that most efficiently describe genealogical relationships and information on divergence (e.g., Farris, 1979).

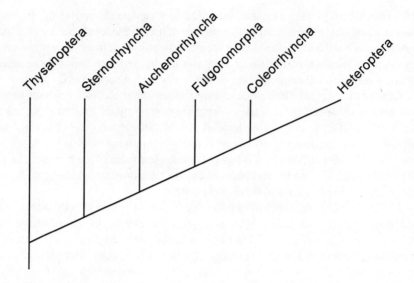

SUBORDINATED:

   Order Thysanoptera
   Order Hemiptera

      Suborder Sternorrhyncha
      Suborder Euhemiptera

         New rank 1 Auchenorrhyncha
         New rank 1 Neohemiptera

            New rank 2 Fulgoromorpha
            New rank 2 Heteropterodea

               New rank 3 Coleorrhyncha
               New rank 3 Heteroptera

SEQUENCED:

   Order Thysanoptera
   Order Hemiptera

      Suborder Sternorrhyncha
      Suborder Auchenorrhyncha
      Suborder Fulgoromorpha
      Suborder Coleorrhyncha
      Suborder Heteroptera

Fig. 8.2. Examples of subordinated and sequenced classifications of the Hemiptera (scale insects, cicadas, true bugs, etc.).

Converting Cladograms to Linnaean Hierarchies

Two approaches have been proposed for converting cladograms into formal classifications. These are *subordination* and *sequencing*. The former approach was advocated by Hennig; the latter approach was apparently first codified by Nelson (1974). These two approaches are demonstrated in Fig. 8.2.

Subordination. This approach requires that each branching level in a cladogram receive a distinctive hierarchic designation. Subordination requires the maximum number of ranks in the Linnaean presentation of the classification, but allows for exact retrieval of the branching pattern without recourse to an additional set of conventions.

Sequencing. This approach allows progressively nested sister-group relationships to be of equal rank, thereby requiring fewer levels, particularly in rendering a pectinate scheme. However, sequencing necessitates knowing the convention by which the Linnaean scheme was rendered in order to retrieve the branching pattern exactly. As shown by Wiley (1979), with enough conventions, any group can be classified or reclassified, without a proliferation of ranks. But the limitation of such an approach is that the conventions must be recorded and remembered.

Classifying Fossils and Taxa of Uncertain Position. The issue of fossil taxa and cladistics was at one time a subject of impassioned debate, primarily because of the traditional paleontological view that fossils might actually represent the ancestors of Recent taxa. We noted in Chapter 4 that there seems to be no way to test theories of ancestor–descendant relationships in a cladistic framework; their admission to the system would allow group recognition based on the absence of characters. Furthermore, as noted by Farris (1976), there is no obvious way to represent ancestor–descendant relationships in a formal hierarchic classification. We might then conclude that all groups — fossil and Recent — should be treated as terminal taxa and that "[t]he main problem of classifying fossils is not the accommodation of 'ancestral groups,' but rather groups of uncertain relationship" (Nelson, 1972:230). Having arrived at this conclusion, the same methods can be used for dealing with all *incertae sedis* taxa — those of uncertain position — irrespective of whether those taxa are Recent or fossil.

Taxa of uncertain position often attain that distinction because they are incomplete, as is the case with many fossil specimens. In a cladogram, such taxa often form part of a trichotomy or polytomy, being related to other taxa only at the level that the available character information allows. A practical approach was advocated by Nelson (1972), whereby fossil groups to which no Recent members are assigned are preceded by a dagger. Taxa of uncertain position are placed at the level at which their relationships can be positively determined and labeled as *incertae sedis*. The following example will illustrate the point:

Class A
  Order A
  †Species A (*incertae sedis*)
    Family A
      Genus A
        Species B
        Species C

> Order B
>   Family B
>   Species D (*incertae sedis*)
>   Genus B
>   Species E
>   Species F

Species A (a fossil) can only be placed with certainty at the level of order, whereas species D (non-fossil) can only be placed with certainty at the level of family. The species B, C, E, and F have sufficient information to place them at the generic level. Further discussion of the classification of fossil taxa can be found in the works of Farris (1976), Patterson and Rosen (1977), and Wiley (1979).

Conclusions. If a consensus has been reached on how to represent the results of phylogenetic analysis in formal classifications, it would seem to embody the following points:

1. Only monophyletic groups should be named.
2. Named groups should be those that need to be recognized by name (i.e., empty ranks need not be named). A good example is the Machilida, discussed earlier in this chapter. As the sister group of all other ectognathous insects, eight ranks would be required to produce a portrayal of ranks for approximately 300 species of bristletails equivalent to that required to deal with the 1 million remaining species of ectognathous insects, namely: Dicondylia, Pterygota, Paleoptera, Paurometabola, Paraneoptera, Acercaria, Condylognatha, and Order. In fact, the two levels most commonly used are Subclass Archaeognatha and Order Machilida.
3. Traditional rankings may be retained (e.g., Class Aves and Class Insecta) with an implicit adoption of the sequencing convention. Hierarchic relationships are established through the naming and (sometimes) ranking of necessary inclusive groupings. The insect orders represent one of the best examples of this practice. Some of the currently recognized groupings were established by Linnaeus and have been used at the same rank level ever since. Phylogenetic studies have revealed, however, that all insect orders do not merit equal rank in a phylogenetic system, as can be seen in the classification of Hennig (Fig. 8.1). However, through the use of approximately eight ranks above the level of order, the hierarchic relationships of the major insect groups can be portrayed, while at the same time retaining the traditionally recognized orders.

Even though classification and phylogeny are viewed as having an inextricable relationship, many published classifications of living organisms are not based on explicit phylogenetic hypotheses. Thus, much of the discussion in this book de-

scribes a goal rather than an achieved result. The desirability of basing classifications on phylogenetic analyses is widely recognized. Our ability to achieve this
goal is, however, severely limited by a variety of factors, including (1) the limited number of practicing systematists; (2) the limited time span during which
appropriate methods have been rigorously applied; and (3) the reality of not yet
having, for many groups, data capable of answering questions about detailed
relationships among the organisms.

Phylogenetic Taxonomy

Classifications should convey information concerning monophyletic groups.
Even though our conception of those groups may change — improve — over time
this should not be taken as a condemnation of classificatory methods, but rather
as an indication of advancement of knowledge. This reasoning has not gone unchallenged, however.

"Phylogenetic taxonomy" has been proposed as a way to "improve" and
thereby supplant the Linnaean hierarchy. The claims for the necessity of such
action can be viewed as resting largely on a restrictive meaning of "definition"
as denoting an "essence," in the Aristotelean sense, with no consideration of the
meaning attached to that term by practicing taxonomists.

Phylogenetic taxonomy, when applied, defines relationships of taxa in terms
of common ancestry. For example:

> Lepidosauria is defined as *Sphenodon* and squamates and all saurians sharing a more
> recent common ancestor with them than with crocodiles and birds.

According to de Queiroz and Gauthier (1990), defining taxa in these terms
allows evolutionary considerations to enter directly into taxonomic definitions
rather than after the fact. Phylogenetic definitions of taxa would — in the view
of its proponents — clarify the distinction between definition and diagnosis. Definitions of taxon names would become ontological statements referring to groups
that are presumed to exist under the central tenet of common descent, independent of our ability to recognize them. According to de Queiroz and Gauthier
(1990:313), taxa recognized in the "phylogenetic taxonomic system" would not
be concepts, but real things — systems deriving their existence from common ancestry relationships among their parts.

The issue of "reality" can easily be questioned, however, by considering an altered topology as a result of revised phylogenetic analysis. Advocates of phylogenetic taxonomy make little mention of this point, perhaps assuming that concepts of relationship will be obvious and neither questioned nor revised. Yet the
"real" groups defined in phylogenetic taxonomy derive their existence from phylogenetic analyses based on attributes thought to be diagnostic for the groups.

Thus, a logical conflict persists in all presentations of the phylogenetic-taxonomic approach, despite the lengthy and impassioned disquisitions of its proponents. As we have noted elsewhere, the "reality" of taxonomic groupings is not a self-evident property. It is, rather, the result of analysis and synthesis of observation. Those observations are represented by nothing more than the attributes (characters) of the organisms themselves. This property, central to our attempts to acquire knowledge of genealogical relationships among organisms, will not disappear through appeals to "evolutionary considerations" on the part of those promoting phylogenetic taxonomy.

Whether phylogenetic taxonomy will provide a stable basis for our taxonomic concepts of higher taxa was addressed by Dominguez and Wheeler (1997). They concluded that information on hierarchic relationships, implicit in the ranked Linnaean system, would be lost in the rankless system of de Queiroz and Gauthier. They further observed that as hypotheses of relationships are revised, taxa named under the system of de Queiroz and Gauthier could change their level of generality radically, from being part of a group to including that same group.

Systematic Databases

With no fewer than 2.5 million species of living organisms, and possibly many more than that, managing classificatory and other information on those organisms is a massive task. If systematic biology were a business or a branch of the government, a database containing information on "the hierarchy of life" would probably have been created long ago. The realization of such a goal still remains a dream. Nonetheless, the potential value of such a database is great.

A database of systematic information might be designed to manage several categories of information. The categories in the following list could all be construed as pertaining to Hennig's conception of a general reference system for biology.

1. Names (including synonymy and homonymy)
2. Hierarchic relations
3. Literature references
4. Biological associations, economic status, etc.
5. Images
6. Character data and analyses
7. Distributional information

A relational database model for managing these categories of information is presented in Figure 8.3. Other similar models have been proposed, and at least one has appeared in the form of commercial software (Colwell, 1996). Colwell's

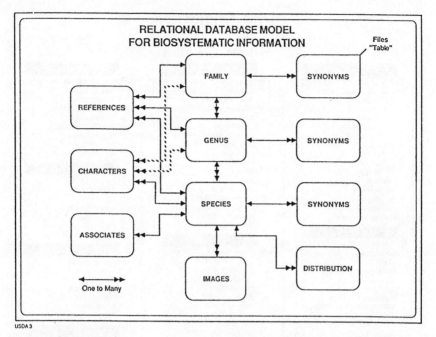

Fig. 8.3. A relational database model for managing information in systematic biology (from Thompson and Knutson, 1987).

Biota model (Fig. 8.4) differs from that of Thompson and Knutson primarily in that it has very limited ability to manage information associated with nomenclature, but has substantial ability to manage locality data directly associated with specimens. These differences can be appreciated by comparing Figures 8.3 and 8.4.

Designing a suitable structure and creating a real-world database are two very different problems. As noted in Chapter 1, the *Index Kewensis* makes basic systematic information on plants available in database format. In addition, a modest number of botanical collections have developed collection databases, and the number will increase.

As is typical, the situation in zoology is much more complex. Modern systematic catalogs (or checklists) do exist for the major vertebrate groups (see Chapter 1). The catalog of McKenna and Bell (1997) for higher groups of mammals was developed as a database, but it is not currently accessible in electronic form.

Literature-based catalogs exist for many invertebrate groups, and the most modern of these are all being produced with the use of some digital technology. The degree to which such works have been compiled as databases, will be converted to databases, or can be accessed in electronic form is highly variable,

Biota Structure

Biota Core Tables

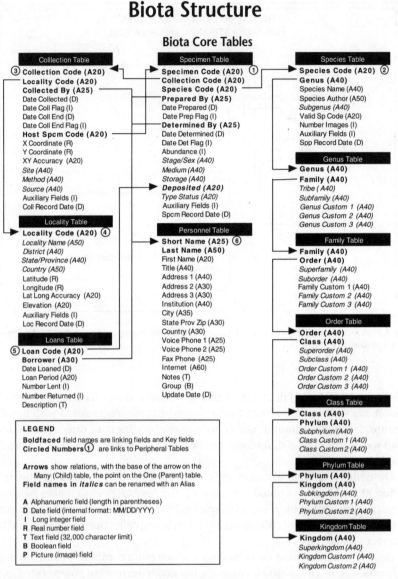

Collection Table
- ③ **Collection Code (A20)**
- **Locality Code (A20)**
- **Collected By (A25)**
- Date Collected (D)
- Date Coll Flag (I)
- Date Coll End (D)
- Date Coll End Flag (I)
- **Host Spcm Code (A20)**
- X Coordinate (R)
- Y Coordinate (R)
- XY Accuracy (A20)
- *Site (A40)*
- *Method (A40)*
- *Source (A40)*
- Auxiliary Fields (I)
- Coll Record Date (D)

Locality Table
- **Locality Code (A20)** ④
- *Locality Name (A50)*
- *District (A40)*
- *State/Province (A40)*
- *Country (A50)*
- Latitude (R)
- Longitude (R)
- Lat Long Accuracy (A20)
- Elevation (A20)
- Auxiliary Fields (I)
- Loc Record Date (D)

Loans Table
- ⑤ **Loan Code (A20)**
- **Borrower (A30)**
- Date Loaned (D)
- Loan Period (A20)
- Number Lent (I)
- Number Returned (I)
- Description (T)

Specimen Table
- **Specimen Code (A20)** ①
- **Collection Code (A20)**
- **Species Code (A20)**
- **Prepared By (A25)**
- Date Prepared (D)
- Date Prep Flag (I)
- **Determined By (A25)**
- Date Determined (D)
- Date Det Flag (I)
- Abundance (I)
- *Stage/Sex (A40)*
- *Medium (A40)*
- *Storage (A40)*
- ***Deposited (A20)***
- *Type Status (A20)*
- Auxiliary Fields (I)
- Spcm Record Date (D)

Personnel Table
- **Short Name (A25)** ⑥
- **Last Name (A50)**
- First Name (A20)
- Title (A40)
- Address 1 (A40)
- Address 2 (A30)
- Address 3 (A30)
- Institution (A40)
- City (A35)
- State Prov Zip (A30)
- Country (A30)
- Voice Phone 1 (A25)
- Voice Phone 2 (A25)
- Fax Phone (A25)
- Internet (A60)
- Notes (T)
- Group (B)
- Update Date (D)

Species Table
- **Species Code (A20)** ②
- **Genus (A40)**
- Species Name (A40)
- Species Author (A50)
- *Subgenus (A40)*
- Valid Sp Code (A20)
- Number Images (I)
- Auxiliary Fields (I)
- Spp Record Date (D)

Genus Table
- **Genus (A40)**
- **Family (A40)**
- *Tribe (A40)*
- *Subfamily (A40)*
- *Genus Custom 1 (A40)*
- *Genus Custom 2 (A40)*
- *Genus Custom 3 (A40)*

Family Table
- **Family (A40)**
- **Order (A40)**
- *Superfamily (A40)*
- *Suborder (A40)*
- Family Custom 1 (A40)
- *Family Custom 2 (A40)*
- *Family Custom 3 (A40)*

Order Table
- **Order (A40)**
- **Class (A40)**
- *Superorder (A40)*
- *Subclass (A40)*
- *Order Custom 1 (A40)*
- *Order Custom 2 (A40)*
- *Order Custom 3 (A40)*

Class Table
- **Class (A40)**
- **Phylum (A40)**
- *Subphylum (A40)*
- *Class Custom 1 (A40)*
- *Class Custom 2 (A40)*

Phylum Table
- **Phylum (A40)**
- **Kingdom (A40)**
- *Subkingdom (A40)*
- *Phylum Custom 1 (A40)*
- *Phylum Custom 2 (A40)*

Kingdom Table
- **Kingdom (A40)**
- *Superkingdom (A40)*
- *Kingdom Custom1 (A40)*
- *Kingdom Custom 2 (A40)*

LEGEND

Boldfaced field names are linking fields and Key fields
Circled Numbers ① are links to Peripheral Tables

Arrows show relations, with the base of the arrow on the
Many (Child) table, the point on the One (Parent) table.
Field names in *italics* can be renamed with an Alias

A Alphanumeric field (length in parentheses)
D Date field (internal format: MM/DD/YYYY)
I Long integer field
R Real number field
T Text field (32,000 character limit)
B Boolean field
P Picture (image) field

Fig. 8.4. Schematic representation of Biota database structure (Colwell, 1996: 521; courtesy of R. K. Colwell).

however, and none rival the completeness or ease of accessibility seen in the *Index Kewensis.*

Databases of vertebrate collections — like those for plants — are relatively straightforward from a logistical point of view because the specimens are large, often have individual numbers associated with them, and the absolute number of specimens is manageable. Databases of some vertebrate collections have been developed.

Insects, and some other nonvertebrate groups, present the greatest problems for specimen-based databases because of the very large numbers of specimens and because individual specimens are seldom labeled so that they can be uniquely identified. Bar coding, used in some plant-specimen databases, has already been utilized to a limited degree for insects and may offer a solution to providing the unique identifiers required for database creation. The use of bar codes is built into the Biota database.

In the long run, the real value of databases will be the following:

1. The ability for continuous updates.
2. The ability to make complex queries, not only within taxa but also across taxa, particularly with regard to geographical distributions and biotic associations.
3. The ability to easily exchange information across networks such as the World Wide Web.

Within the near future we can expect to see many additional electronic compilations of collection- and literature-based systematic information. Advances in and increased availability of database and other digital technology will undoubtedly affect the form of such works, and facilitate access to the information they contain.

Literature Cited

Ashlock, P. D. 1979. An evolutionary systematist's view of classification. *Syst. Zool.* 441–450.

Colwell, R. K. 1996. *Biota: The Biodiversity Database Manager. Software and Manual.* Sinauer Associates Inc., Sunderland, Massachusetts.

de Queiroz, K., and J. Gauthier. 1990. Phylogeny as a central principle in taxonomy: phylogenetic definitions of taxon names. *Syst. Zool.* 39:307–322.

Dominguez, E., and Q. D. Wheeler. 1997. Taxonomic stability is ignorance. *Cladistics* 13:367–372.

Farris, J. S. 1976. Phylogenetic classification of fossils with Recent species. *Syst. Zool.* 25:271–282.

Farris, J. S. 1979. The information content of the phylogenetic system. *Syst. Zool.* 28:483–519.

Gaffney, E. S. 1979. An introduction to the logic of phylogeny reconstruction. pp. 79–111. *In:* Cracraft, J., and N. Eldredge (eds.), *Phylogenetic Analysis and Paleontology.* Columbia University Press, New York.

Hennig, W. 1969. *Die Stammesgeschichte der Insekten.* Waldemar Kramer, Frankfurt am Mein. 436 pp.

Knox, E. B. 1998. The use of hierarchies as organization models in systematics. *Biol. J. Linn. Soc.* 63:1–49.

Mayr, E. 1974. Cladistic analysis or cladistic classification? *Zeit. zool. Syst. Evol.-forsch.* 12:94–128.

McKenna, M. C., and S. K. Bell. 1997. *Classification of Mammals above the Species Level.* Columbia University Press, New York. 535 pp.

Nelson, G. 1972. Phylogenetic relationship and classification. *Syst. Zool.* 21:227–231.

Nelson, G. 1974 (1973). Classification as an expression of phylogenetic relationships. *Syst. Zool.* 22:344–359.

Patterson, C., and D. E. Rosen. 1977. Review of ichthyodectiform and other Mesozoic teleost fishes and the theory and practice of classifying fossils. *Bull. Amer. Mus. Nat. Hist.* 158:1–172.

Platnick, N. I. 1978. Classifications, historical narratives, and hypotheses. *Syst. Zool.* 27:365–369.

Thompson, F. C., and L. Knutson. 1987. Catalogues, checklists and lists: a need for some definitions, new words, and ideas. *Antenna* 11:131–134.

Wiley, E. O. 1979. An annotated Linnaean hierarchy, with comments on natural taxa and competing systems. *Syst. Zool.* 28:308–337.

Suggested Readings

Farris, J. S. 1979. The information content of the phylogenetic system. *Syst. Zool.* 28:483–519. [A sometimes technical discussion of how and why hierarchic classifications store and transmit information]

Wiley, E. O. 1979. An annotated Linnaean hierarchy, with comments on natural taxa and competing systems. *Syst. Zool.* 28:308–337. [A detailed discussion of classificatory conventions]

9

Historical Biogeography and Host–Parasite Co-evolution

In Chapter 7 we saw how consensus techniques can be used as a way of discovering information in common among multiple most parsimonious clado-grams and for forming classifications. We will now examine the comparison of topological relationships among cladograms as a method for understanding his-torical biogeographic relationships and co-evolutionary patterns between para-sites and their hosts.

Co-evolution is used in this chapter to refer to those cases where hosts and their parasites appear to have intimate, long-standing historical connections. This type of association obtains for many internal parasites and for certain exter-nal parasites, such as lice. Nonetheless, there is a whole class of host–parasite relationships where the association between the parasite group and the host does not show such long-term fidelity, but involves many apparent host shifts. The lat-ter type of association will be addressed in Chapter 10.

Historical Biogeography

A Brief Historical Review

Whereas methods for discovering the ordered distribution of attributes among organisms are now well established, the same cannot be said for tech-niques used to analyze repetitive patterns of geographic distribution. Nelson (1978:269) noted that "Biogeography is a strange discipline. In general, there are no institutes of biogeography; There are no departments of it. There are no professional biogeographers — no professors of it, no curators of it." We will re-view very briefly the history and nature of biogeographic inquiry from the late 1850s forward, paying particular attention to relationship of cladistics and biotic distributions. Reviews of the pre-Darwinian history of biogeography are given by Nelson (1978) and Nelson and Platnick (1981: Chapter 6).

The term *biogeography* has had many meanings attached to it. We will treat it as having two connotations: (1) the study of biotic distributions and their short-term ecological influences, a field often referred to as "geographical ecology" or "ecological geography"; and (2) the study of biotic distributions resulting from longer-term historical factors, an area of inquiry usually called "historical biogeography." Treatises on the subject of biotic distributions have not always made clear these distinctions, and indeed some authors have purposely blurred them, as for example Endler (1982). Biogeography as used in the following pages means *historical biogeography*.

The study of historical biogeography traditionally has been heavily influenced by the state of knowledge of geology and geologic processes. Thus, when the orthodoxy of geology did not allow for continental movements, biogeographers usually attributed distributional disjunctions solely to the dispersal powers of the organisms involved. When plate tectonics became the accepted explanation for orogeny, ocean formation, and other feature changes on the earth's surface, biogeographers soon began to adapt their explanations to the new geological paradigm.

Some biogeographers have proposed distributional theories independent of the influence from other fields of inquiry. For example, the Darwinian contemporary Joseph Dalton Hooker, one of the most influential botanists of his time, was, and remained, an ardent supporter of the idea that the biota of New Zealand, Australia, and South America achieved its present distribution via previous land connections among the continents. Likemindedly, in one of his early works on biotic distributions in the Malaya Archipelago, Alfred Russell Wallace (1860) argued for the former connection of land masses as a way of explaining his observations on animal distributions in the area. Whereas Hooker remained true to what he believed to be the most obvious explanation of the data (i.e., land connections), Wallace soon capitulated to the Darwinian — dispersalist — point of view, which was virulently against modification of geological features as a method for explaining observed biotic distributions (Fichman, 1977). The stabilist view of geography adopted by Darwin and many of his contemporaries derived from their aversion to the "fanciful" creations such as land bridges invoked by Hooker and others to explain modern-day continental biotic disjunctions involving tremendous distances. They believed dispersal of the organisms themselves, across the oceans or circuitously over the contiguous land masses, offered a more reasonable explanation for these observed phenomena.

But, after reading the early-twentieth-century expositions of Alfred Wegener (e.g., Wegener, 1966) on continental drift, you will likely conclude that the data — even at that time — spoke clearly and strongly for the concept of continental movements. For example, Wegener's arguments included geodetic data about the movement of Greenland relative to Europe; an analysis of the excellent fits of the continental outlines of North America and Europe and South America and Africa; among others, the Paleozoic geological similarities of the Cape Province of South Africa and the mountains in Buenos Aires Province, Argentina;

and the known distribution of evidence for Permo-Carboniferous glaciation on the southern continents showing the "invalidity of [continental] permanence theory." Wegener also found substantial corroboration of his theory in the distributional patterns of Recent plants and animals. In short, there seemed to Wegener to be a massive body of corroborative evidence for the theory of continental movement as opposed to the then-current theories of a contracting earth, subsiding land bridges, or stable continents.

Wegener's views were received enthusiastically by some, including the South African geologist–paleontologist Alexander du Toit. However, influential Americans, including mammalogist William Diller Matthew, geologist–paleontologist George Gaylord Simpson, and entomologist–biogeographer Philip J. Darlington were particularly resolute in rejecting Wegener's ideas as a way of explaining biotic distributions. This, in the case of Darlington and Simpson, even after the acquisition of modern geophysical evidence for continental movements.

One might wonder whether the evidence of biotic distributions has a legitimate voice of its own, or if it must always be tempered with the more "credible" knowledge of earth processes derived from studies in geology. History suggests that plants and animals do have a story to tell, and those who have disregarded the biotic evidence have often failed to appreciate the ultimate strength of the data provided by organisms themselves. One need only compare the conclusions of Darlington (1965) and Brundin (1966) with regard to the biotas of the far Southern Hemisphere. Darlington rejected cladistics and also resisted accepting continental drift as an explanation of the distributions of those biotas. He "felt certain" that all plants and animals reached New Zealand across the water, at least during the Tertiary and probably before that. Darlington maintained this viewpoint even for those many plant groups with distributions restricted to the southern portions of the southern continents and with connections among those land masses. In a great leap of logic, he then concluded (Darlington, 1965:107):

> If these plants and all other plants and animals that have reached New Zealand crossed water gaps, no land connections are needed anywhere across the southern end of the world to explain the distribution of far-southern terrestrial life.

Brundin, through the use of phylogenetic methods, demonstrated repeated patterns of intercontinental connection within lineages of chironomid midges in the far south (see Fig. 9.4). He attributed those phylogenetic connections to former land connections, including Antarctica, which he explained via continental drift. In short, whereas Darlington adhered to the prevailing systematic and geological dogma of his formative years and rejected explaining easily observed patterns with the bolder tools of plate tectonics and cladistics, Brundin accepted those tools and treated better-documented patterns of geographic distribution as the result of common causes.

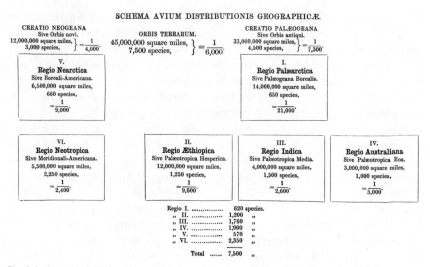

Fig. 9.1. Sclater's (1858) scheme of biotic regions, based primarily on the study of birds.

Having admitted that there is a geographical pattern to the distribution of plants and animals, how might we analyze it? The early history of this enterprise was largely descriptive and probably best characterized in the work of the British ornithologist Philip Sclater. Sclater's mark on the field was made in his paper "On the General Geographical Distribution of the Members of the Class Aves" (Sclater, 1858), in which he proposed six biogeographic "regions." Sclater's division of the world's biota roughly corresponded to the major continental land masses, with the exception of the Oriental Region (Regio Indica) (Fig. 9.1), and was soon accepted by Wallace, in his works on biogeography (Wallace, 1876), and by many other workers. Sclater's regions represented recognizable geographic features that showed major biotic differences among themselves. It would be some time before Leon Croizat, a botanist, and Willi Hennig recognized that distributional data might be subject to new and improved means of analysis, irrespective of the geological orthodoxy. The works of both authors proved formidable to comprehend.

Croizat's works are prolix in the extreme, achieving a level of verbosity possibly unequalled in the biological literature. His major ideas are most mercifully summarized in *Space, Time, Form: The Biological Synthesis* (Croizat, 1962). Croizat said, in reference to common patterns of animal and plant distributions across the continents, "Nature forever repeats," an observation that he believed demanded explanation. He dubbed his approach "Panbiogeography," the core activity of which was "track analysis." The data on relationships and distributions of taxa came from monographs and revisions, much of it from works that had nothing to do with the Euphorbiacaeae, the group of plants on which Croi-

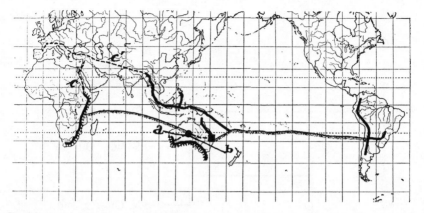

Fig. 9.2. Tracks indicating the distribution of the angiosperm family Proteaceae (Personioideae and Grevilloideae) (from Croizat, 1962: 169).

zat himself had specialized. Tracks were drawn by Croizat to connect disjunct areas possessing related organisms, such as members of the angiosperm family Proteaceae, as shown in Fig. 9.2. This observed phenomenon was termed "vicariance." Its explanation was geologic history. The approach of Croizat has been applied in its original form by only a few workers, notably Craw (e.g., 1982, 1983), although the seeming reality of many of the repetitive distributions identified by Croizat has been documented by many. An example comes from the work of Weston and Crisp (1994) on the waratahs, a subgroup of Proteaceae (Fig. 9.3), which when analyzed in detail, portray a pattern of distribution very similar to that depicted by Croizat.

Hennig's work on biogeography was described by Ashlock (1974) as "at once brilliant and badly conceived [and] its title and organization are such that few but dipterists would be attracted." Hennig's major contribution to biogeographic analysis (Hennig, 1960) dealt with the Diptera of New Zealand as an example problem and was an extension of his methodological work on phylogenetic relationships. Hennig's approach relied heavily on the *progression rule*— the idea that if an organism has migrated to new areas, developing new characters as its distribution expands, the progress of migration is marked by sequentially more-derived characters.

Hennig's work on biogeography, despite what are now viewed as its flaws, is notable because it established a direct connection between phylogenetic analysis and the study of biotic distributions. Nonetheless, it might have remained obscure, even in translation, had it not been for the monographic work of the Swedish entomologist Lars Brundin (1966), who applied both Hennig's phylogenetic and biogeographic methodologies to his studies of chironomid midges from the southern continents. Brundin showed that there were repetitive dis-

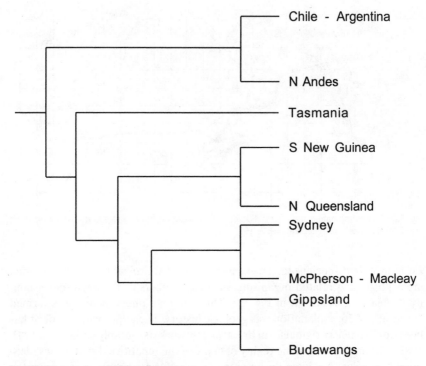

Fig. 9.3. Intercontinental connections between eastern Australia and montane and southern South America as shown by the waratahs, a subgroup of Proteaceae (from Weston and Crisp, 1994).

tributional patterns among midges of that part of the world (Fig. 9.4) and that — in vindication of the views of Joseph Dalton Hooker with regard to plants — the midge distributions connected the southern continents. Furthermore, in Brundin's view, the patterns supported Hennig's progression rule. The significance of Brundin's findings was widely recognized. The appearance of his work coincided felicitously with the spreading belief that continental drift was a scientifically credible theory and with the publication of two readily accessible English language renderings of Hennig's phylogenetic methodology (Hennig, 1965, 1966).

The empirical work of Rosen (1975, 1978) and the theoretical work of Nelson and Platnick (Platnick and Nelson, 1978; Nelson and Platnick, 1981; and Platnick, 1981) laid the groundwork for a more mature biogeographic method. It is the approach developed by these authors — sometimes called *cladistic biogeography,* as by Humphries and Parenti (1986) — that has come to dominate the historical study of plant and animal distributions. The analytic tools of the field are still being refined. Thus, the following presentation should be interpreted as the state of a developing approach to analysis.

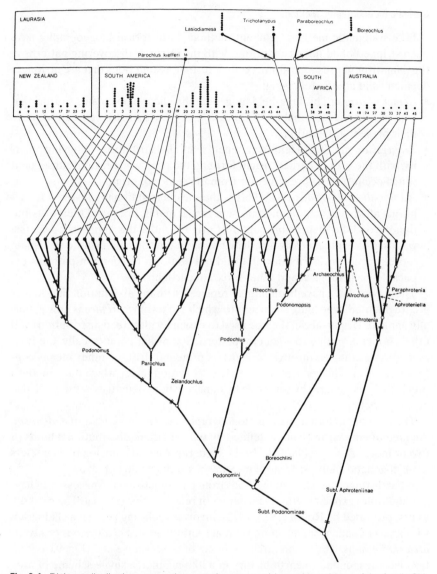

Fig. 9.4. Disjunct distributions on southern continents as evidenced by midges of the family Chironomidae as portrayed by Brundin (Nelson and Ladiges, 1996: fig. 2; courtesy of The American Museum of Natural History). Note the repeated pattern of occurrence in New Zealand, South America, and Australia in separate groups of midges.

Cladistic Biogeographic Methods

Like systematic methods, current methods in historical biogeography have been stripped of ad hoc assumptions, with the intent of discovering patterns in the data without presupposition. Among former assumptions that are no longer deemed valid are the following:

- *Center of origin:* This long-held biogeographic dictum postulates a priori that taxa arise in one identifiable area and *disperse* from there to eventually achieve their present distribution.
- *Progression rule:* This Hennigian tenet treats all distributions as having polarity, with the taxa of earliest origin forming the root, while taxa of subsequent origin occur progressively further from the center of origin.
- *Ad hoc theories of dispersal:* This is the means traditionally invoked for allowing all taxa to achieve their present range; that is, to expand their distribution beyond the identifiable center of origin. Under this approach, dispersal was used to explain nearly any distribution, rather than being used to explain those aspects of distribution that did not conform to common patterns explicable via geological and other forces of earth history.

These postulates have been largely replaced with the assumption that ranges are fragmented over time, a concept to which the term *vicariance* is now generally applied. Dispersal can then be used to explain widespread distributions and others not conforming to a general pattern. Somewhat paradoxically, the fragmentation of ranges, through orogeny, continental drift, or other means conforms closely to the concept of allopatric speciation under which most modern precladistic biogeography was done by noncladistic systematists — using the dispersal model.

The data of modern historical biogeography are *taxa* and *areas of endemism.* An area of endemism "can be defined by the congruent distributional limits of two or more species" (Platnick, 1991). Taxa represent the analog of characters in phylogenetic analysis; areas of endemism are the analog of taxa.

What distinguishes current methods from past practice is the necessity of having (multiple) explicit hierarchic schemes of relationships for taxa. The method, as first proposed by Rosen (1975, 1978), involves replacing taxa on a cladogram with areas of endemism, forming what are known as *area cladograms* or *taxon-area cladograms.* Thus, a hierarchic scheme of areas is revealed (Fig. 9.6). If all taxa have responded to earth history in a like manner, a single scheme of area interrelationships — a *general area cladogram* — would be the result, and no further analysis would be required. Observation suggests, however, that such is seldom, if ever, the case for a variety of reasons that might include the following:

1. Taxa may be of different ages and therefore may have vicariated (or otherwise differentiated) in response to different phenomena.
2. Different taxa may have responded to the same phenomenon in different ways.

3. Our theories concerning areas of endemism may be in error because of subsequent dispersal, extinction, poor sampling, or the hybridization of land areas.

The criterion for postulating areas of endemism is simply the known geographic occurrence of a taxon. The test of the theory — that other taxa occur there — is usually much less clear-cut than is the case with morphological characters (see Harold and Mooi, 1994).

Nonetheless, there are guidelines, possibly even methods, for refining our concepts of areas of endemism. Platnick (1991) observed that we might always wish to pursue biogeographic studies using taxa with the smallest ranges and the largest numbers of species for the area under study. One of his favored examples is the treatment of cool-temperate southern South America as a single area, even though many groups of spiders, and some other chelicerate arthropods such as Opiliones, show extreme diversity and highly localized distributions in this geographic region.

Sidebar 11
Biogeographic Terminology

The following terms are widely used in the discussion of cladistic biogeography. Clarification of their intended meaning will help the reader to comprehend the remainder of the discussion in this chapter.

- *component* — a (monophyletic) group of taxa (areas) connected at a node on a cladogram.
- *general area cladogram* — a cladogram showing the resolution of area relationships among a variety of taxon-area cladograms.
- *missing taxon* — a taxon missing from one of the areas in one of the area cladograms being compared, the result being noncorrespondence of distributions across taxa.
- *redundant (paralogous) distributions* — the sequential or multiple occurrence of the same area on an area cladogram; a phenomenon observed to occur frequently among the basal taxa (areas) on cladograms (Nelson and Ladiges, 1996).
- *taxon-area cladogram* (or *area cladogram*) — a cladogram of taxa for which the distributions of the taxa have been substituted for the taxa themselves.
- *term* — a terminal taxon (area) on a cladogram.
- *three-area statement* — the application of three taxon analysis to areas of endemism.
- *widespread taxon* — a taxon that occurs in more than one area of endemism as areas of endemism are understood on the basis of other taxa.

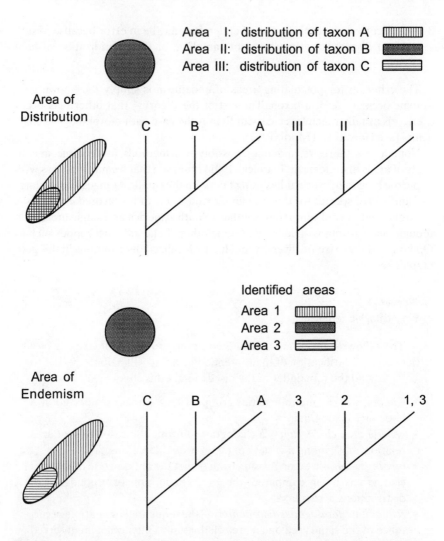

Fig. 9.5. Determining areas of endemism as potentially different from areas of distribution, according to the method of Axelius (1991). The upper portion of the figure shows distributions coded as areas. The bottom portion shows areas of endemism coded to recognize overlapping areas as distinct for all taxa.

The treatment of the known distributions of species as de facto areas of endemism was criticized by Axelius (1991) in cases where the distributions of species are overlapping. He offered the solution shown in Fig. 9.5. The argument for this approach lies in the observation that the overlap in the distributions of taxa A and B must be due to dispersal of one or both of the species. Therefore, the portion of the distribution of taxon A overlapping with the distribution of taxon

B should be coded separately from the nonoverlapping portion. Otherwise, the resulting area cladograms will not be completely informative with regard to the areas occupied by the taxa. In cases of partial overlap, Axelius recommended coding the overlapping portion as a separate area, in the anticipation of determining which part of the distribution was the result of dispersal.

The core problem of cladistic biogeography then becomes, "How do we combine nonidentical area cladograms for different groups to produce a summary of the information common to them?" The process proceeds as follows (Nelson and Platnick, 1981; Page, 1990a; Morrone and Carpenter, 1994):

1a. Construction of taxon-area cladograms from taxon cladograms. This step is heavily observational in nature. Areas of endemism are determined for all taxa, the widespread taxa being coded as occupying more than one area. The taxon cladogram is then labeled with the areas of endemism, as shown in Fig. 9.6.

1b. Conversion of taxon-area cladograms into resolved area cladograms. As originally described in detail by Nelson and Platnick (1981) in their exposition of "component analysis," any given area cladogram may imply a few to many "resolved area cladograms," primarily because of the occurrence of widespread taxa. Missing areas will result in some resolved area cladograms being different, but will generally not increase the numbers of resolved area cladograms.

Four approaches have been proposed for producing resolved area cladograms, these representing a set of *assumptions* — with arcane names — about how to deal with widespread taxa and missing areas. The solutions range from very restrictive to quite liberal. A novel solution for paralogous (redundant) distributions has been offered by Nelson and Ladiges (1996), and will be discussed below.

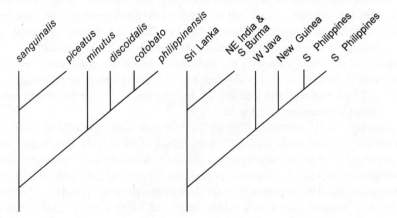

Fig. 9.6. Taxon cladogram (Stonedahl, 1988) and corresponding area cladogram (Schuh and Stonedahl, 1986).

First, let us examine the attributes of the four assumptions which can be summarized as in Table 9.1 (modified from Page, 1990a):

Table 9.1. Four Assumptions for Forming Resolved Area Cladograms

	Problem	
	Missing areas	Widespread taxa
Brooks Parsimony Analysis (BPA):	uninformative	sister areas (monophyletic)
Assumption 0:	primitively absent	sister areas (monophyletic)
Assumption 1:	uninformative	paraphyletic
Assumption 2:	uninformative	float all but one occurrence

Brooks Parsimony Analysis (BPA) (Wiley, 1988) derives directly from methods proposed by Brooks (1981) for the analysis of host–parasite data, where additive binary coding methods are used to prepare a matrix of data from area cladograms, which can then be analyzed using a standard parsimony program. BPA and Assumption 0 (Zandee and Roos, 1987) use similar methods for forming resolved area cladograms. They both treat the areas occupied by widespread taxa as monophyletic (i.e., the areas that combine to form a widespread distribution are more closely related to one another than to any other areas occupied by the taxa under consideration). This aspect of these two "assumptions" makes them the most restrictive of the four. The Brooks Parsimony Analysis and Assumption 0 differ only in their treatment of missing areas, which are viewed as uninformative by BPA and primitively absent under Assumption 0.

Assumption 1 (Nelson and Platnick, 1981) is less restrictive in that area relationships for widespread taxa can be either monophyletic or paraphyletic with respect to the taxon inhabiting them (Fig. 9.7). Thus, the widespread taxon may be informative in combination with other taxon-area cladograms if the components of the cladogram with the widespread taxon are combinable with the components in a taxon-area cladogram where the widespread distribution is resolved. The taxon with the widespread distribution will never be recognized as representing more than one taxon.

Assumption 2 (Nelson and Platnick, 1981) is least restrictive with regard to the treatment of widespread taxa, and in the minds of some is the most reasonable interpretation of many biogeographic data sets. A widespread taxon on a single cladogram comprises a component containing unresolved area relationships, but when considered in concert with other cladograms the unresolved area relationships may become informative (Humphries and Parenti, 1986). Under this interpretation, some — but not all — of the areas of endemism composing the distribution of widespread taxa may "float" on the general area cladogram. Thus,

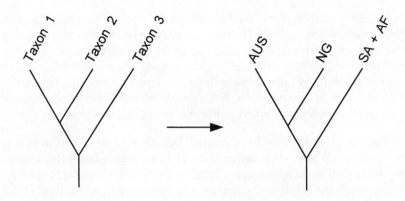

Assumption 1

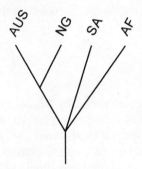

Assumption 2

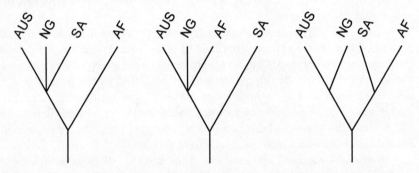

Fig. 9.7. Taxon-area cladogram and resolved area cladograms under Assumption 1 and Assumption 2. AUS = Australia; NG = New Guinea; SA = South America; AF = Africa. Assumption 1 does not require that SA + AF be monophyletic, as would be the case with Brooks Parsimony Analysis and Assumption 0, but it does not specify how either one might be related to the remaining areas. Assumption 2 specifies those possible area relationships (from Humphries and Parenti, 1986).

what were uninterpretable area relationships on the single taxon-area cladogram become resolved (see Fig. 9.7). The taxon with the widespread distribution may in the future be interpreted as representing more than one taxon.

As pointed out by Page (1990a), these four need not be the only assumptions one might make about the resolution of distributional information in order to prepare resolved area cladograms. For example, all possible combinations of ways of treating missing areas and widespread taxa could be invoked.

2. Preparation of general area cladograms. Determining what information is shared in common across taxa can be achieved in a variety of ways. Both parsimony and consensus techniques have been used for finding general area cladograms; some of the possible approaches have been described by Page (1988, 1989, 1990a). As pointed out by Morrone and Carpenter (1994) in their comparison of automated methods, two areas are still causing confusion: (1) the choice between parsimony and consensus is not clear because neither method produced results that Morrone and Carpenter found conclusive; and (2) the available software could only solve problems involving a small number of taxa and areas.

Subtree Analysis and Area Paralogy

Area paralogy (redundant distributions) has been identified by Nelson and Ladiges (1996) as the source of apparent inconsistency in biogeographic data as seen, for example, in the results of Morrone and Carpenter cited above. The removal of paralogy, through what Nelson and Ladiges termed "subtree analysis," appears to offer improved results. Improved, in the sense used here, implies (1) the discovery of greater congruence among taxon-area cladograms, (2) the ability to solve problems involving more than ten taxa and areas (including widespread distributions), and (3) the ability to analyze these larger data sets in a reasonable period of time.

Nelson and Ladiges (1996) coded matrices for their data in the form of components, following the approach outlined by Nelson and Platnick (1981), and as three-area statements (see Chapter 6). The results were analyzed using standard parsimony programs; the unsupported nodes produced by the programs were deleted through a node-by-node examination of the resultant general area cladograms. The preparation of subtrees, the rationale for which is depicted in Fig. 9.8, and the analysis of the data with a parsimony program are responsible for the efficiency with which the results were produced.

Nelson and Ladiges (1996) established the utility of their approach by reanalyzing most of the data sets dealt with by Morrone and Carpenter, which allowed direct comparability between the results of the methods applied by the two groups of investigators. They also analyzed several large data sets, including those of Brundin (1966), Schuh and Stonedahl (1986), and Mayden (1988). Whereas Morrone and Carpenter often found multiple solutions, sometimes

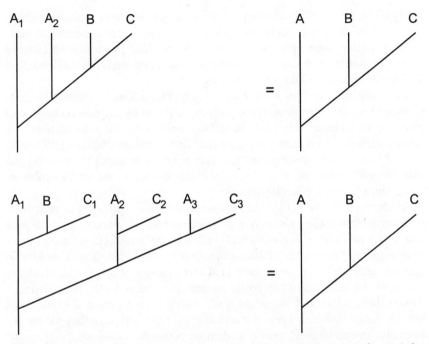

Fig. 9.8. Two examples of taxon-area cladograms with paralogous distributions, before and after subtree analysis. Subscripts represent the repetitive occurrence of different taxa in the same area (from Nelson and Ladiges, 1996).

thousands, Nelson and Ladiges found one to a few solutions for the same data sets. The latter authors concluded that removal of paralogy from area clado-grams allows for solutions to complex problems and that the number of possible solutions is greatly reduced. The "assumptions" described above are no longer needed with subtree analysis, as ambiguity among distributions is removed by other means.

Criticism of Cladistic Biogeography

Cladistic biogeography is by its very nature concerned with forming connec-tions among biotas with disjunct distributions. Craw (1982, 1983) criticized "cla-distic" biogeography for being preoccupied with fragmentation and therefore incapable of providing a reliable analysis of the biotas of already composite areas. Polhemus (1996:63), without citing Craw, echoed the same sentiments: "Cladistic and vicariance biogeographers . . . developed methods that due to their reliance on dichotomously branching diagrams were best suited to the depiction of faunal patterns arising from continental fragmentation . . . these methods have been notably less successful when applied to areas of continental convergence [and that] Present component and parsimony analyses that rely solely on dichotomous branching . . . are clearly insufficient tools for such investigations." Both authors suggested that a single land area appearing in more than one pattern would represent failure of the method, presumably because of its being based on dichotomy and fragmentation. They used New Zealand, New Caledonia, and New Guinea as examples.

The complaints of Craw were interpreted by Platnick and Nelson (1984:331) as suggesting that "the geographic pattern exhibited by the preponderance of evidence would be accepted as real, and that conflicting real patterns shown by other taxa would be rejected as unreal (i.e., due to chance dispersal)." They indicated that no such assumptions had ever been made under the method and that, indeed, in the case of geologically composite areas, we would expect incongruence between general patterns.

Schuh and Stonedahl (1986) had shown New Guinea appearing in two places on a taxon-area cladogram, in one case relating New Guinea to tropical Asia and in the other case to continental Australia. Polhemus (1996) used this as an example of "the futility of attempting to search for a single area cladogram that will accurately represent the faunal relationships of an area in which arc collisions and subsequent composite terranes have been dominant influences." Nonetheless, Schuh and Stonedahl made it clear that the method would force the conclusion that New Guinea was of hybrid origin, although they did not try to resolve the problem of areas of endemism within New Guinea.

Host–Parasite Co-evolution

The observation that hosts and their parasites have evolved in concert is long-standing, and was discussed extensively by Hennig (1966) under the heading of "Fahrenholz's rule." The development of improved techniques for determining to what degree "co-evolution" has actually occurred, however, has gone hand-in-hand with the development of biogeographic methods.

Hennig (1966:112) believed that the relationships of parasites to their hosts contained much information of great value to phylogenetic systematics, but lamented the unsatisfactory state of its theoretic basis. He asserted that "Even the most extreme advocates of the thesis that the phylogeny of the parasites usually parallels the phylogeny of the host (in the sense of Fahrenholz's rule) do not assume that the parallelism is so close that every process of speciation in the one corresponds to a process of speciation in the other" (Hennig, 1966:111). This statement suggests a parallel with historical biogeography.

Host–parasite comparisons benefit from the fact that both hosts and parasites have phylogenies constructed from intrinsic character data. Thus the problem of "homologous taxa" does not arise, as it does with areas of endemism. The similarities between host–parasite relationships and biotic distributions might be listed as follows:

1. Some hosts may have no parasites, the biogeographic equivalent of certain taxa being absent from certain areas.
2. Some parasites may occur on more than one host, the biogeographic equivalent of widespread distributions.
3. Some hosts have more than one parasite, the biogeographic equivalent of redundant distributions.

Brooks (1981) explicitly treated the analysis of phylogenetic relationships of multiple parasites on a single group of hosts as a way of establishing (testing) the relationships of the hosts themselves. He used the parasite relationships as character data and assumed that with multiple parasite phylogenics signal would overwhelm noise and a corroborated answer would result, showing a correct host cladogram based only on cladograms of parasites. Most subsequent authors have not followed Brook's approach and have avoided a priori assumptions about co-evolution. Rather, they have more frequently used consensus techniques as a way of determining to what degree parasites have actually evolved in parallel with their hosts.

As a first example of such congruence-based studies, let us examine the work of Farrell and Mitter (1990) on leaf beetles of the genus *Phyllobrotica* (Coleoptera: Chrysomelidae) and their hosts in the closely related flowering plant families Lamiaceae and Verbenaceae. These authors constructed a cladogram for the 14 species of *Phyllobrotica* and two outgroup taxa. They also constructed a scheme of relationships for the hosts, based on the best estimates available in the literature. They then compared the topologies of the host and parasite cladograms, excluding from the comparison those five beetle species for which no host information was available. Farrell and Mitter calculated the consensus and the Adams compromise for the two cladograms. They found the Adams result superior to the strict consensus for comparisons of this type because it allowed that "both parallel phylogenesis and host transfer may contribute to a given set of insect/plant interactions." In the Adams result, six of the eight possible groupings were resolved, suggesting a high degree of parallel diversification between the lamialean hosts and chrysomelid parasites.

As a second example let us consider work on co-evolution between pocket gophers and their louse parasites as originally conducted by Hafner and Nadler (1988; 1990) (Fig. 9.9). The data were reanalyzed by Page (1990b, 1994), with substantial observations on methods for seeking congruence between host and parasite phylogenies. The objects of this study were eight species of pocket gophers in the genera *Geomys, Orthogeomys,* and *Thomomys* (Rodentia: Geomyidae). They are the hosts of 10 species of chewing lice belonging to the genera *Geomydoecus* and *Thomomydoecus* (Insecta: Phthiraptera). Hafner and Nadler identified six components common to the two phylogenies and postulated three instances of dispersal to explain the remaining louse distributions. Page reanalyzed the data of Hafner and Nadler to determine whether those au-

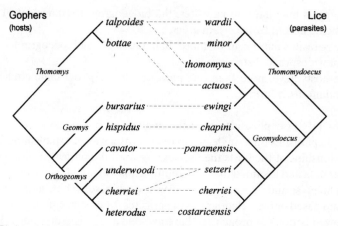

Fig. 9.9. Phylogenetic relationships of pocket gophers and those of their louse parasites compared. Dotted lines indicate relationships of the hosts and their parasites (from Page, 1994: fig. 9).

thors had identified the maximum number of co-speciation events. He found that they had, but concluded that the anomalous distributions could be explained alternatively as the result of sampling error and extinction. Under further analysis, Page concluded that rather than two possible reconstructions of relationships for the pocket gophers and their lice, there were actually six. The maximum number of co-speciation events was always six; the differences among the schemes concerned alternative explanations for incongruence between the lice and their hosts.

Literature Cited

Ashlock, P. D. 1974. The uses of cladistics. *Ann. Rev. Ecol. Syst.* 5:81–99.

Axelius, B. 1991. Areas of distribution and areas of endemism. *Cladistics* 7:197–199.

Brooks, D. R. 1981. Hennig's parasitological method: a proposed solution. *Syst. Zool.* 30:229–249.

Brundin, L. 1966. Transantarctic relationships and their significance, as evidenced by chironomid midges. *Kungl. Svenska Vetenskapsakademiens Handlingar.* Fjarde series, 11:1–472.

Craw, R. C. 1982. Phylogenetics, areas, geology, and the biogeography of Croizat: a radical view. *Syst. Zool.* 31:304–316.

Craw, R. C. 1983. Panbiogeography and vicariance cladistics: Are they truly different? *Syst. Zool.* 32:431–438.

Croizat, L. 1962. *Space, Time, Form: The Biological Synthesis.* Published by the author, Caracas. 881 pp.

Darlington, P. J. 1965. *Biogeography of the Southern End of the World: Distribution and history of far-southern life and land, with an assessment of continental drift.* McGraw-Hill, New York. 236 pp.

Endler, J. A. 1982. Problems of distinguishing historical from ecological factors in bio-geography. *Amer. Zool.* 22:441–452.

Farrell, B., and C. Mitter. 1990. Phylogenesis of insect/plant interactions: Have *Phyllobrotica* leaf beetles (Chrysomelidae) and the Lamiales diversified in parallel? *Evolution* 44:1389–1403.

Fichman, M. 1977. Wallace: zoogeography and the problem of land bridges. *J. Hist. Biol.* 10:45–63.

Hafner, M. S., and S. A. Nadler. 1988. Phylogenetic trees support the coevolution of parasites and their hosts. *Nature* 332:258–259.

Hafner, M. S., and S. A. Nadler. 1990. Co-speciation in host-parasite assemblages: comparative analysis of rates of evolution and timing of cospeciation. *Syst. Zool.* 39:192–204.

Harold, A. S., and R. D. Mooi. 1994. Areas of endemism: definition and recognition criteria. *Syst. Biol.* 43:261–266.

Hennig, W. 1960. Die Dipteren-Fauna von Neuseeland als systematische und tiergeographisches Problem. *Beitr. Entomol.* 10:221–329. [English translation by: Wygodzinsky, P. 1966. The Diptera fauna of New Zealand as a problem in systematics and biogeography. *Pacific Insects Monograph* 9:1–81].

Hennig, W. 1965. Phylogenetic Systematics. *Ann. Rev. Entomol.* 10:97–116.

Hennig, W. 1966. *Phylogenetic Systematics*. University of Illinois Press, Urbana. 263 pp.

Humphries, C. J., and L. R. Parenti. 1986. *Cladistic Biogeography*. Clarendon Press, Oxford. 98 pp.

Mayden, R. L. 1988. Vicariance biogeography, parsimony, and evolution in North American freshwater fishes. *Syst. Zool.* 37:329–355.

Morrone, J. J., and J. M. Carpenter. 1994. In search of a method for cladistic biogeography: an empirical comparison of component analysis, Brooks parsimony analysis, and three-area statements. *Cladistics* 10:99–153.

Nelson, G. 1978. From Candolle to Croizat: comments on the history of biogeography. *J. Hist. Biol.* 11:269–305.

Nelson, G., and P. Y. Ladiges. 1996. Paralogy in cladistic biogeography and analysis of paralogy-free subtrees. *Amer. Mus. Novitates* 3167:44 pp.

Nelson, G., and N. Platnick. 1981. *Systematics and Biogeography: Cladistics and Vicariance*. Columbia University Press, New York. 567 pp.

Page, R. D. M. 1988. Quantitative cladistic biogeography: constructing and comparing area cladograms. *Syst. Zool.* 37:254–270.

Page, R. D. M. 1989. Comments on component compatibility in historical biogeography. *Cladistics* 5:167–182.

Page, R. D. M. 1990a. Component analysis: a valiant failure? *Cladistics* 6:119–136.

Page, R. D. M. 1990b. Temporal congruence and cladistic analysis of biogeography and cospeciation. *Syst. Zool.* 39:205–226.

Page, R. D. M. 1994. Parallel phylogenies: reconstructing the history of host-parasite assemblages. *Cladistics* 10:155–173.

Platnick, N. I. 1981. Widespread taxa and biogeographic congruence. pp. 223–227. *In:* Funk, V. A., and D. R. Brooks (eds.), *Advances in Cladistics: Proceedings of the First Meeting of the Willi Hennig Society*. New York Botanical Garden, Bronx, New York.

Platnick, N. I. 1991. On areas of endemism. *In:* Ladiges, P. Y., C. J. Humphries, and L. W. Martinelli (eds.). *Austral Biogeography*. CSIRO, Canberra, Australia.

Platnick, N. I., and G. Nelson. 1978. A method of analysis for historical biogeography. *Syst. Zool.* 27:1–16.

Platnick, N. I., and G. Nelson. 1984. Composite areas in vicariance biogeography. *Syst. Zool.* 33:328–335.

Polhemus, D. A. 1996. Island arcs, and their influence on Indo-Pacific biogeography. pp. 51–66. *In:* Keast, A., and S. E. Miller (eds.), *The Origin and Evolution of Pacific Island Biotas, New Guinea to Eastern Polynesia: Patterns and Processes.* SPB Academic Publishing, Amsterdam.

Rosen, D. E. 1975. A vicariance model of Caribbean biogeography. *Syst. Zool.* 24:431–464.

Rosen, D. E. 1978. Vicariant patterns and historical explanation in biogeography. *Syst. Zool.* 27:159–188.

Schuh, R. T., and G. M. Stonedahl. 1986. Historical biogeography in the Indo-Pacific: a cladistic approach. *Cladistics* 2:337–355.

Sclater, P. L. 1858. On the general geographical distribution of the members of the class Aves. *J. Linn. Soc., Zool.* 2:130–145.

Stonedahl, G. M. 1988. Revisions of *Dioclerus, Harpedona, Mertila, Myiocapsus, Prodromus,* and *Thaumastomiris* (Heteroptera: Miridae, Bryocorinae: Eccritotarsini). *Bull. Amer. Mus. Nat. Hist.* 187:1–99.

Wallace, A. R. 1860. On the zoological geography of the Malay Archipelago. *J. Linn. Soc. London* 4:172–184.

Wallace, A. R. 1876. *The Geographical Distribution of Animals.* McMillan, London.

Wegener, A. 1966. *The Origin of Continents and Oceans.* Dover, New York. 246 pp. [English translation of the fourth edition, 1929]

Weston, P. H., and M. D. Crisp. 1994. Cladistic biogeography of waratahs (Proteaceae: Embothrieae) and their allies across the Pacific. *Aust. Syst. Bot.* 7:225–249.

Wiley, E. O. 1988. Parsimony analysis and vicariance biogeography. *Syst. Zool.* 37:271–290.

Zandee, M., and M. C. Roos. 1987. Component-compatibility in historical biogeography. *Cladistics* 3:305–332.

Suggested Readings

Cracraft, J. 1988. Deep history biogeography: Retrieving the historical pattern of evolving continental biotas. *Syst. Zool.* 37:221–236. [A succinct commentary on biogeography methodology]

Humphries, C. J., and L. R. Parenti. 1986. *Cladistic Biogeography.* Clarendon Press, Oxford. 98 pp. [A concise but thorough treatment of background and method]

Mitter, C., and D. R. Brooks. 1983. Phylogenetic aspects of coevolution. pp. 65–98. *In:* Futuyma, D. J., and M. Slatkin (eds.), *Coevolution.* Sinauer Associates, Sunderland, Massachusetts. [A classic paper on the analysis of coevolutionary relationships]

Morrone, J. J., and J. V. Crisci. 1995. Historical biogeography: introduction to methods. *Ann. Rev. Ecol. Syst.* 26:373–401. [A review of methods in historical biogeography]

Nelson, G., and N. Platnick. 1981. *Systematics and Biogeography: Cladistics and Vicariance.* Columbia University Press, New York. 567 pp. [Historical review and detailed discussion of biogeographic methods]

Nelson, G., and D. E. Rosen (eds.). 1981. *Vicariance Biogeography. A Critique.* Columbia University Press, New York. 593 pp. [A collection of symposium papers, including discussion, dealing broadly with historical biogeography]

10

Ecology, Adaptation, and Evolutionary Scenarios

In Chapter 9 we examined congruence among cladogram topologies as a method for studying degree of historical correlation. That method of testing does not usually form a satisfactory basis for testing theories of ecological associations, adaptational hypotheses, and more loosely constrained host associations. It is, nonetheless, desirable to evaluate such theories in a historical context. Indeed, this area of inquiry has been the subject of several books (see Suggested Readings). Two interrelated analytic approaches have been proposed — mapping and optimization. We will review examples of the application of each.

Evolutionary Scenarios and Their Tests

The acquisition of attributes and associations over evolutionary time has long fascinated biologists. Views on the how, when, and why of such acquisitions have often been expressed in the form of scenarios. For example, arachnologists long held the view that the orb web evolved twice from a cobweblike structure because of its superior prey-catching ability (Coddington, 1988). Although such statements might be viewed as hypotheses to be tested, they have frequently been propounded as if test was irrelevant or as if their truth content was self-evident. Whatever the intention of the authors of such statements, many such scenarios are subject to test via knowledge derived from systematic studies. In other words, using systematics we might come to some understanding of whether an idea represents mere assertion or whether available data actually support it.

Corroboration, or testing, of such scenarios might be divided into two broad approaches: (1) *cladistic methods,* which provide a way of understanding the sequence of ecological and adaptive changes; or (2) *statistical techniques,* which are used for testing theories of adaptation while uncorroborated cladograms or nonphylogenetic classifications are uncritically used as a standard of comparison (Crisp, 1994). We will examine the former approach in some detail. The latter approach, whose base of inquiry is rooted in ecology rather than in systematics, has been promoted under the banner of "comparative biology" by Harvey and Pagel (1991) and will not be covered here.

Mapping

Once a cladogram is available, the relationships of attributes among the terminal taxa can be understood explicitly. Thus, ecological associations and other forms of data not used in the phylogenetic analysis can be "mapped" onto a cladogram as a way of understanding the number of origins, the sequence of origin, and the presumed ancestral conditions.

Mapping was used by Mitter, Farrell, and Wiegmann (1988) to test the widely invoked — but seldom tested — hypothesis that "diversification is accelerated by adoption of a new way of life, i.e., movement into a new adaptive zone." Mitter et al. mapped the feeding habits of major clades of insects onto cladograms of those groups and compared the numbers of known species in adjacent clades. Their results showed that in 11 of 13 comparisons, the phytophagous lineage was significantly more diverse than its nonphytophagous sister group.

Some monophyletic groups of phytophages are known to maintain strict associations with a particular host group. This is not the case with many groups, however, nor does it obtain at higher levels of relationship. Because there is no logical criterion available for determining homology and state order for such extrinsic characters, mapping is the only way to understand the sequence of change. Miller (1987) used mapping to demonstrate that patterns of host association in the swallow-tail butterflies of the family Papilionidae were often not tightly constrained at the tribal or generic level (Fig. 10.1). Nonetheless, he concluded, on the basis of the patterns of evidence, that host switching in the Papilionidae is probably constrained by plant chemical compounds.

Mapping was used at the species level by Futuyma and McCafferty (1990) in their examination of host relationships of the chrysomelid beetle genus *Ophraella* and its association with members of the plant family Asteraceae. Although the beetles are restricted to one plant family, critical analysis showed no clear patterns of co-speciation between the beetles and the eight genera of Asteraceae that serve as their hosts. The authors concluded that host shifts by the beetles postdate the divergence of the host plants.

Mapping has also been used to analyze the historical origins of traits that are at least in part genetically determined. As one possible example, Andersen (1997) studied the evolution of flightlessness and wing polymorphism in insects generally, the suborder Heteroptera, and the family Gerridae in particular. Andersen found that patterns were difficult to discern at the level of insect orders, probably because the nature of the condition is difficult to define. Within the Gerridae, however, his analysis produced the "unexpected" conclusion that — contrary to popular conception — the ancestral condition for the wings is not always monomorphic macroptery (fully developed functional wings). Andersen's analysis revealed that within the Gerridae wing dimorphism (fully winged

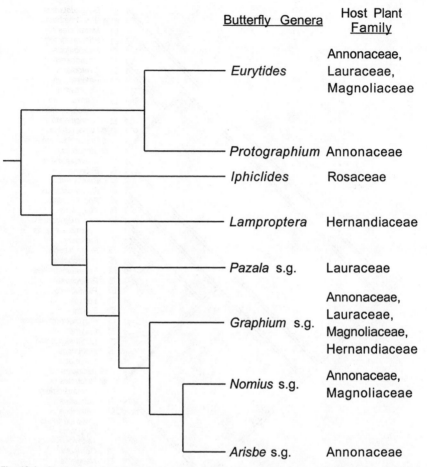

Fig. 10.1. Cladogram depicting phylogenetic relationships among genera of the tribe Graphiini (Papilionidae) compared with host plant associations in the group (Miller, 1987). Comparison shows that not only do different butterfly lineages feed on the same host group, but also that some butterfly lineages feed on multiple host groups. Thus, we must assume multiple colonizations or losses of association with some plant groups.

forms and those with reduced wings and flight musculature) is primitive, with monomorphic macroptery being a secondary acquisition. As can be seen from Fig. 10.2, Andersen's analysis permits determination of the ancestral condition for nodes on the cladogram, as would be attempted under the "optimization" approach described below.

In another case, Wenzel (1993) examined congruence between behavior and morphology. His study group comprised 23 of the 28 generic-level taxa of the paper wasp subfamily Polistinae. Wenzel treated the attributes of nest architecture as behavioral characters. Although we have argued that behavioral characters can be treated as intrinsic and therefore used as part of a "total evidence"

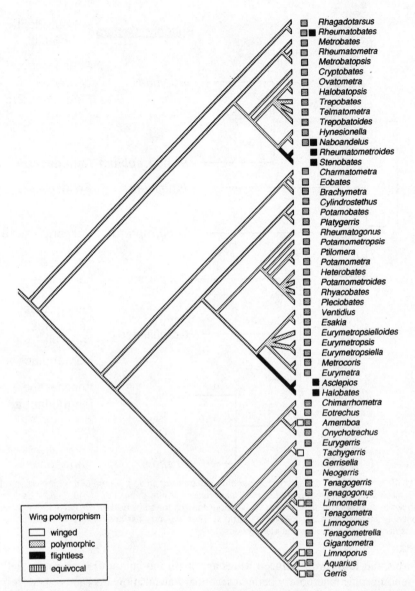

Fig. 10.2. Mapping flightlessness on a cladogram of generic relationships within the family Gerridae (Andersen, 1997:98), showing the widespread and apparently primitive nature of wing polymorphism, and the relative rarity of occurrence and fixity of flightlessness and permanent macroptery (fully developed functional wings).

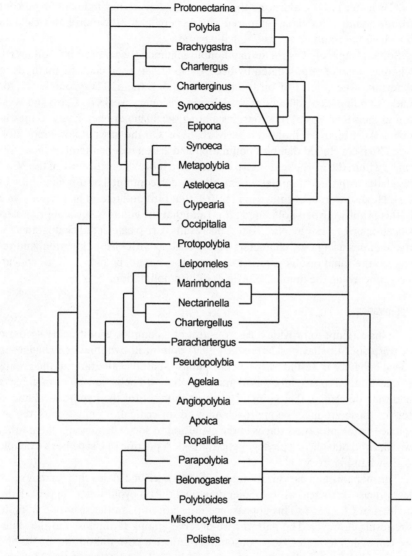

Protonectarina
Polybia
Brachygastra
Chartergus
Charterginus
Synoecoides
Epipona
Synoeca
Metapolybia
Asteloeca
Clypearia
Occipitalia
Protopolybia
Leipomeles
Marimbonda
Nectarinella
Chartergellus
Parachartergus
Pseudopolybia
Agelaia
Angiopolybia
Apoica
Ropalidia
Parapolybia
Belonogaster
Polybioides
Mischocyttarus
Polistes

Morphology   Architecture

Fig. 10.3. Comparison of cladograms based on morphology (*left*) and nest architecture (*right*) for the paper wasp subfamily Polistinae (from Wenzel, 1993).

analysis, Wenzel chose to construct a cladogram solely on the basis of behavioral traits in order to examine the degree to which those attributes produced a scheme congruent with one based on morphology. His comparison of the two schemes is shown in Fig. 10.3, revealing a substantial degree of congruence. He also optimized his "architectural" data directly onto the morphology-based cladogram of Carpenter (1991), a process that revealed a consistency index of 56 percent for the architectural data, compared with 48 percent for the morphological data on which the cladogram was originally based.

Some groups of Australian papilionoid legumes (peas) are bird pollinated, whereas others are pollinated by bees. Crisp (1994) used cladistic methods to determine the origins of bird pollination among them. The groundwork for understanding bird pollination was laid in revisionary works by Crisp and Weston in their analyses of relationships among the 30 genera belonging to the endemic Australian papilionoid tribe Mirbelieae. On the basis of floral morphology, Crisp concluded that bird pollination had arisen independently at least five times within the Mirbelieae. Crisp conducted a more detailed study of the *Nemcia* clade, with three nominal genera, all of which contained bird-pollinated species. He discovered, as shown in Fig. 10.4, that bird pollination had arisen twice in this primitively bee-pollinated group and that the genus *Nemcia* of prior classifications was paraphyletic. To test his results, Crisp calculated cladograms for the legumes using only characters not associated with bird pollination, and retrieved the same results, offering evidence for the independence of his conclusion concerning the number of origins of bird pollination.

Optimization

In his attempt to provide a rigorous test of a complex evolutionary scenario of wasp social behavior, Carpenter (1989) optimized hypothesized changes in social behavior in vespid wasps on morphology-based cladograms of the groups in question. His procedure was to first map observed behaviors on terminal taxa and then determine the optimal character-state set for the hypothetical ancestors of those groups. The approach allowed for critical examination of the sequence of the proposed stages in the evolution of social behavior, a theory that was derived outside of a phylogenetic context. A portion of Carpenter's example is shown in Fig. 10.5.

The mapping and optimization approaches described in this chapter may seem less precise or to deal with sloppier data than the co-evolutionary approach described in Chapter 9. This appearance does not impugn the scientific value of the result, however. The patterns we observe among terminals, and the states we can assign to nodes, offer substantial predictive power about the habits and associations for members of groups yet to be collected or observed. This predictive power is the same as for attributes whose existence derives directly from inheritance.

Test of Adaptational Hypotheses

From the selectionist view of evolution come theories predicated on the idea that selection for superior adaptations acts as the force for evolutionary change.

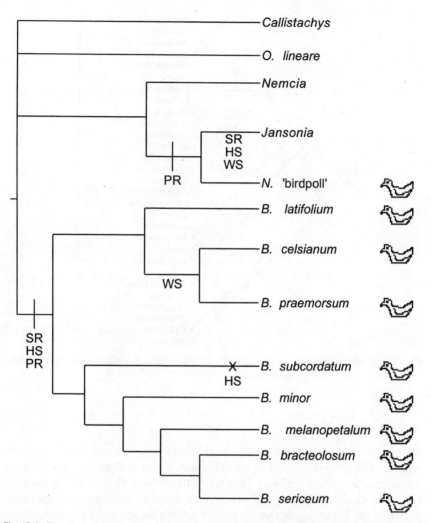

Fig. 10.4. The cladistic analysis of Crisp (1994), showing two independent origins of bird pollination in the Australian papilionoid legume tribe Mirbelieae and the paraphyletic nature of the genus *Nemcia*. *B.* = Brachysema; SR = standard reduced, shorter than keel; HS = habit scandent; PR = petals red; WS = wings shorter than keel. The bird symbols indicate the presence of bird pollination in a group.

Many such theories have not been subjected to rigorous testing, but exist more in the realm of ideas. Coddington (1988) proposed an approach utilizing cladistic methods to test such theories. The requirements of his approach, portrayed in Fig. 10.6, demand that cladograms be constructed for the group being studied on the basis of attributes other than those involved in the hypothesis of

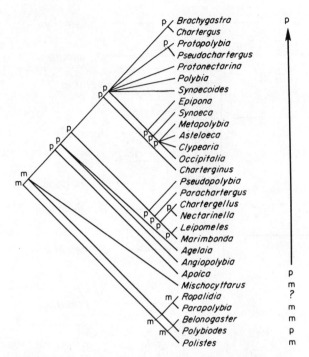

Fig. 10.5. Optimizing changes on the nodes of a cladogram of polistine vespid wasps. m = short-term monogyny; p = swarm-founding polygyny (Carpenter, 1989).

adaptation. Also necessary is a specified relationship between form (morphology, M) and associated adaptation (function, F), as well as the stipulation of transformation from primitive (ancestral) to derived. One of the hypotheses tested by Coddington was that orb webs evolved twice from cobwebs (mentioned at the beginning of this chapter). Coddington showed that orb webs are ancient and had only evolved once and that cobwebs are advanced relative to orb webs, not vice versa.

The mapping and optimization approaches described above allow us to structure our understanding of extrinsic attributes in light of the best available theories of phylogenetic relationships. These extrinsic data may or may not have been construed under some prior theory of relationships for the taxa exhibiting them. Nonetheless, the type of adaptational theories discussed by Coddington imply theories of relationships. Such theories will survive testing only if they conform to the requirements specified by Coddington and also conform to the independently derived phylogenetic theory.

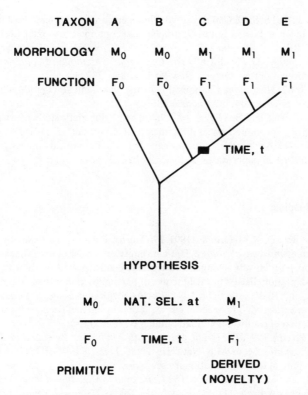

Fig. 10.6. The approach of Coddington (1988: fig. 1) for testing hypotheses of evolution through adaptation. M = morphology; F = function; subscripts indicate direction of hypothesized change.

Literature Cited

Andersen, N. M. 1997. Phylogenetic tests of evolutionary scenarios: the evolution of flightlessness and wing polymorphism in insects. pp. 91–108. *In:* Grandcolas, P. (ed.), *The Origin of Biodiversity in Insects: Phylogenetic Tests of Evolutionary Scenarios.* Mem. Mus. Natn. Hist. Nat., Paris, vol. 173.

Carpenter, J. M. 1989. Testing scenarios: wasp social behavior. *Cladistics* 5:131–144.

Carpenter, J. M. 1991. Phylogenetic relationships and the origin of social behavior in the Vespidae, pp. 7–32. *In:* Ross, K. G., and R. W. Matthews (eds.), *Social Biology of Wasps.* Cornell University Press, Ithaca, New York.

Coddington, J. A. 1988. Cladistic tests of adaptational hypotheses. *Cladistics* 4:3–22.

Crisp, M. D. 1994. Evolution of bird-pollination in some Australian legumes (Fabaceae). pp. 281–309. *In:* Eggleton, P., and R. Vane-Wright (eds.), *Phylogenetics and Ecology.* Linnaean Society Symposium Series, no. 17. Academic Press, London and San Diego. 376 pp.

Futuyma, D. J., and S. S. McCafferty. 1990. Phylogeny and evolution of host plant associations in the leaf beetle genus *Ophraella* (Coleoptera: Chrysomelidae). *Evolution* 44:1885–1913.

Harvey, P. H., and M. D. Pagel. 1991. *The Comparative Method in Evolutionary Biology.* Oxford University Press, Oxford, New York, Tokyo. 239 pp.

Miller, J. S. 1987. Host-plant relationships in the Papilionidae (Lepidoptera): parallel cladogenesis or colonization? *Cladistics* 3:105–120.

Mitter, C., B. Farrell, and B. Wiegmann. 1988. The phylogenetic study of adaptive zones: has phytophagy promoted insect diversification? *Amer. Nat.* 132:107–128.

Wenzel, J. W. 1993. Application of the biogenetic law to behavioral ontogeny: a test using nest architecture in paper wasps. *J. Evol. Biol.* 6:229–247.

Suggested Readings

Brooks, D. R., and D. A. McLennan. 1991. *Phylogeny, Ecology, and Behavior. A Research Program in Comparative Biology.* The University of Chicago Press, Chicago and London. 434 pp. [A relatively strong introduction to cladistics, with analytic tools for ecology and adaptation, biased toward the use of parsimony rather than congruence]

Eggleton, P., and R. Vane-Wright. 1994. *Phylogenetics and Ecology.* Linnaean Society Symposium Series. No. 17. Academic Press, London and San Diego. 376 pp. [Papers on the application of cladistics in understanding ecological issues]

Grandcolas, P. (ed.). 1997. *The Origin of Biodiversity in Insects: Phylogenetic Tests of Evolutionary Scenarios.* Mem. Mus. Natn. Hist. Nat., Paris, vol. 173. [Discussions of methods and empirical tests of evolutionary scenarios]

Harvey, P. H., and M. D. Pagel. 1991. *The Comparative Method in Evolutionary Biology.* Oxford University Press, Oxford, New York, Tokyo. 239 pp. [A relatively weak treatment of cladistics, with analytic tools tending toward the statistical]

11

Biodiversity and Conservation

Increasing awareness of the diminishment of natural habitats on a global scale has brought the study of "biodiversity" to a new level of intensity. *Biodiversity,* as the term is currently used, has many meanings, and its study ranges broadly across biology. There are, however, aspects of biodiversity that are strictly systematic, including (1) enumeration of the world's biota and (2) determination of historical relations — both genealogical and geographical — among members of the biota. This information can be used directly to inform our efforts for staving off continuing extinction at the hand of the human race. It is these areas that will be the subject of the following pages.

Enumeration

Linnaeus, the eminent eighteenth-century Swedish naturalist, described the world biota between 1740 and 1767, at least as he knew it. We now know that a complete enumeration of all living species may never be achieved, no matter the number of workers performing the task. Nonetheless, we should not give up on the effort because much of importance remains to be known. We might mention, among other possibilities, the following areas impacted by improved knowledge of the world's biota: the functioning of ecosystems, human health, agriculture and other aspects of food production, discovery of natural products, and conservation decision making.

Estimates for the total number of species of living organisms range from 2.5 million to about 30 million, depending on who is doing the calculations and what criteria they apply. Whatever the number, one thing is certain: large numbers of species remain unstudied, even at the most basic level of describing and naming them so as to formally recognize their existence.

Certain aspects of current knowledge of the world's biota may in some ways seem counterintuitive. For example, the insect fauna of Europe is extremely well known. The description of a new species might be considered an important event. By comparison, the insect fauna of North America is not nearly so well known.

Not that it might not be, given the potentially immense resources available for science in the United States. Yet, large numbers of species remain undescribed, and some groups, particularly in the western United States, have never been studied in detail.

Even with the existing limitations of knowledge of the North American fauna, the temperate Northern Hemisphere is extremely well known compared to the tropics and the temperate Southern Hemisphere. The biota of Australia may be least well known, but other areas are in desperate need of study as well. An example may provide some perspective on the scale of the problem. Within the insect family Miridae, 2000 species have been described from North America north of Mexico (Henry and Wheeler, 1988). One might surmise that when the North American fauna is "completely" known, that number might increase by as much as 15 percent. Australia, with a comparable land area, has a currently described fauna of 180 species (Cassis and Gross, 1995). The number of species actually existing in Australia is certainly not 180, however. The authors of the *Insects of Australia* (CSIRO, 1991) estimated the number at 600, although even this number is much too small, judging from specimens available in collections. One might conclude that, whereas approximately 15 percent of the North American species remain to be described, no more than 8 percent of the Australian species have been described. These numbers could be repeated for many groups of insects, other arthropods, and other nonvertebrate animals.

The situation for most groups of insects might also be compared with that for flowering plants. In North America the discovery of new species of angiosperms would probably be considered newsworthy. There are several floras for North America and often multiple floras for the various "regions" within North America. Australia, by comparison, as yet has no complete flora and large numbers of species of flowering plants remain undescribed.

Historical Relations

Genealogical

Without the relatively complete enumeration of species, knowledge of phylogenetics will always be limited. Yet, theories of genealogical relationships provide systematists with their most powerful tools. To believe that credible genealogies are known for most groups of plants and animals would be naive at best, and probably better described as delusional. Furthermore, there is no obvious correlation between the degree to which species of a group are known, and the level of knowledge concerning phylogenetic relationships of those species and the higher groups to which they belong.

For example, the species of mammals and birds are quite well understood. It is relatively rare that new taxa are described. Phylogenetic relationships among those taxa are not as well known as one might expect, however. We might say that ornithologists were distracted by the "new systematics" for most of half a century, during which time they devoted their efforts to examining population-level variation within birds, almost to the exclusion of studying higher-level relationships (but see Sibley and Ahlquist [1990] on DNA hybridization). Relationships among mammals may be better understood; at the very minimum, the structure of relationships as currently understood at the generic level and above has been portrayed in detail by McKenna and Bell (1997).

These examples represent only a glimpse into where things stand. What seems clear is that without a substantially greater expenditure of effort, many species will never be identified as existing, let alone studied at the level of phylogenetic relationships.

Geographical

Knowledge of historical biogeographic relationships is dependent on phylogenetic knowledge. The significance of the relationship between knowledge of genealogy and historical biogeography at a global level may be best appreciated by examining some recent observations of Platnick (1992). Platnick noted that traditionally theories of historical biogeographic relationships among animals have been dominated by studies of vertebrates. This domination can be portrayed by observing that a onetime widely used textbook of "zoogeography" by the entomologist Philip J. Darlington (1957) used vertebrates as the sole example organisms. Platnick further noted that studies of zoogeography have also been heavily weighted toward the Northern Hemisphere, creating what he referred to as the "boreal megafaunal bias." He emphasized that it is not just the tropics that contain large numbers of taxa compared to the Northern Hemisphere — as popular conception would have it — but that the southern continents in general deserve much greater recognition as centers of great biotic diversity and therefore biogeographic importance. On the latter point, he stressed that areas of endemism in the far south often seem to be much smaller than in the Northern Hemisphere.

We might profitably extend Platnick's discussion to the issue of regional phylogenetic diversity. But first, let us augment some of Platnick's numbers. We can agree that the numbers of species in some areas of the Southern Hemisphere are virtually unparalleled anywhere in the world. As Platnick noted, the South African flora is made up of 15,000–16,000 species of flowering plants (Acocks, 1953). Western Australia should also be cited in this context, with an estimated 4000 species of flowering plants occurring in the Southwestern Botanical Province alone (Corrick, Fuhrer, and George, 1996), those species comprising at least 20 percent of the Australian angiosperm species, but occupying roughly 4 percent of the continental land area. North America probably has no more than 9000 species of flowering plants, by comparison, with no single area of the continent showing such conspicuously high diversity.

We might also wish to consider Southern Hemisphere diversity in plant-feeding insects. Data compiled by Zimmerman (1991–1994) for the weevils (Insecta: Coleoptera: Curculionidae) of Australia indicate an estimated 6000 to 8000 species in 1000 genera (a very large proportion remaining to be described), whereas the North American fauna, by comparison, comprises 2500 species in 375 genera. Another example can be found in the work of Slater on the monocot-feeding lygaeoid family Blissidae (Heteroptera). The South African fauna comprises 66 species placed in 11 genera, these feeding on the monocot families Poaceae, Cyperaceae, Haemodoraceae, Juncaceae, and Restionaceae (Slater and Wilcox, 1973). The North American fauna, by comparison, contains only 26 species in 2 genera, these all feeding on the Poaceae (Slater, 1964). Yet, the land area of South Africa is less than 13 percent of that of North America. All of these observations offer support for the idea of higher diversity and smaller areas of endemism in the southern continents.

Southern floristic connections in groups such as the Proteaceae were well known in the time of Hooker and Darwin. However, it was the Swedish entomologist Brundin (1966) who first showed through the use of phylogenetic methods the detailed nature of the massive connections existing between the southern continents, exclusive of the northern biota. Whereas the type of pattern exemplified by Brundin's examples from the midges of the family Chironomidae involved the cool-temperate portions of the Southern Hemisphere — presumably including Antarctica — tropical patterns also exist. For example, the plant bug tribe Pilophorini was shown by Schuh (1991) to demonstrate biotic connections between tropical South America, tropical Africa, and tropical Asia in its relatively more primitive lineages, whereas only the relatively most recent lineages occur in the Northern Hemisphere, albeit with a large number of species.

Discrete distributions and large numbers of clades often distinguish the biotas of the southern continents. By contrast the fauna of the Holarctic is often much more homogeneous, with the genera and species having relatively broad distributions. In a paper on biogeographic relationships within the Holarctic, Enghoff (1995), divided the Northern Hemisphere into only four areas: Eastern and Western Nearctic and Eastern and Western Palearctic. No matter how poorly this division reflects the actual number of areas of endemism, it nonetheless graphically reflects the degree to which many taxa have widespread distributions.

Conservation

How, then, can phylogenetic and biogeographic information be used in conservation decision-making?

Phylogenetic Information

Two nearly diametrically opposite approaches have been suggested, with a number of variants in between (e.g., Nixon and Wheeler, 1992), for using phylogenetic information in conservation decision making.

Protect radiating lineages. Under this approach, lineages that appear to be radiating (speciating) on a dynamic basis would be those targeted for protection because they are the ones that would "create future biodiversity" (Erwin, 1991). The extreme application of this approach would be to treat lineages that are not radiating, and therefore "doomed to go extinct," as expendable and unworthy of efforts at protection.

Protect the most lineages. Under the "opposite" approach, absolute numbers of species would be deemed less important than the lineages that those species compose (Vane-Wright, Humphries, and Williams, 1991). Thus, ancient and species-poor phyletic lines are deemed more important than recent and species-rich lineages.

Example. The application of the latter approach can be appreciated by using the example of the tuatara of New Zealand — the sole surviving species (either one or two) of an otherwise long extinct reptilian lineage (see May, 1990). If two species are actually represented, as taxonomic studies suggest, then it might be worth a concerted effort to protect them both because of their phylogenetic "uniqueness." If, on the other hand, such a conservation question involved one of many species of murine rodents (the largest group of mammals), the answer would not be couched in terms of phylogenetic uniqueness, and the former approach of protecting radiating lineages might be applied.

Biogeographic Information

Conservation decisions often involve determining if a given area should be protected or developed, irrespective of what organisms might or might not occur there. In such cases, information on *endemicity* must usually be assessed before any phylogenetic information is taken into account. The concept of *surrogacy* has taken on special significance in the biodiversity–conservation literature because making complete biotic samples is time consuming and therefore usually not feasible. Surrogacy is the practice of choosing certain groups of organisms to be representative of the biota as a whole. To what degree surrogacy works may well be determined by the quality of indicator organisms chosen in each case under study. Knowledge derived from the study of biogeography would suggest that the concept must be used with care.

Consider, for example, comparisons between flowering plants and insects.

Australia may have slightly more than twice as many angiosperm as weevil species. It would be much easier to sample the plant species for a given area than the weevils. Yet, nothing is known about the degree of correspondence between plant and weevil endemicity. Even if such a correlation were assumed, we still might not choose plants as the sole surrogate group because weevils may not be uniformly distributed across flowering plants as hosts. Furthermore, phylogenetically unique groups might not occur in areas of highest endemism.

A very coarse-scale approach to examining diversity was taken by Williams, Humphries, and Gaston (1994). In this case study, the families of seed plants were used as surrogates for an overall measure of species richness. The ultimate value of the approach might be tested by determining to what degree it measures diversity when compared with results of applying the same higher-taxon approach to other groups of organisms.

The above arguments provide a glimpse into the ways systematic information can be used in our efforts to protect and preserve the world's endangered biota. The refinement of these approaches is a process in an active state of discussion and development.

Literature Cited

Acocks, J. P. H. 1953. *Veld Types of South Africa.* Botanical Survey Memoir No. 28. Department of Agriculture, Pretoria. 192 pp.

Brundin, L. 1966. Transantarctic relationships and their significance, as evidenced by chironomid midges. *Kungl. Svenska Vetenskapsakademiens Handlingar.* Fjarde series, 11:1–472.

Cassis, G., and G. F. Gross. 1995. *Zoological Catalog of Australia.* Vol. 27.3A. Hemiptera: Heteroptera (Coleorrhyncha to Cimicomorpha). Australian Biological Resources Study, CSIRO, Melbourne. 506 pp.

Corrick, M. G., B. A. Fuhrer, and A. S. George. 1996. *Wildflowers of Southern Western Australia.* Five Mile Press, Noble Park. 224 pp.

CSIRO. 1991. Hemiptera. *In: The Insects of Australia,* vol. 1., 429–509. 2 vols. Melbourne University Press.

Darlington, P. J., Jr. 1957. *Zoogeography: The Geographical Distribution of Animals.* John Wiley and Sons, New York. 675 pp.

Enghoff, H. 1995. Historical biogeography of the Holarctic: area relationships, ancestral areas, and dispersal of non-marine animals. *Cladistics* 11:223–263.

Erwin, T. L. 1991. An evolutionary basis for conservation strategies. *Science* 253:750–752.

Henry, T. J., and A. G. Wheeler, Jr. 1988. Family Miridae. *In: Catalog of Heteroptera, or True Bugs, of Canada and the Continental United States.* E. J. Brill, Leiden. 958 pp.

May, R. M. 1990. Taxonomy as destiny. *Nature* 129–130.

McKenna, M. C., and S. K. Bell. 1997. *Classification of Mammals above the Species Level.* Columbia University Press, New York. 535 pp. [Part I: History and Theory of Classification]

Nixon, K. C., and Q. D. Wheeler. 1992. Measures of phylogenetic diversity. pp. 216–234.

In: Novacek, M. J., and Q. D. Wheeler (eds.), *Extinction and Phylogeny.* Columbia University Press, New York.

Platnick, N. I. 1992. Patterns of biodiversity. pp. 15–24. *In:* Eldredge, N. (ed.), *Systematics, Ecology, and the Biodiversity Crisis.* Columbia University Press, New York. 220 pp.

Schuh, R. T. 1991. Phylogenetic, host, and biogeographic analyses of the Pilophorini (Heteroptera: Miridae: Phylinae). *Cladistics* 7:157–189.

Sibley, C. G., and J. E. Ahlquist. 1990. *Phylogeny and Classification of Birds: A Study in Molecular Evolution.* Yale University Press, New Haven, Connecticut.

Slater, J. A. 1964. *A Catalogue of the Lygaeidae of the World.* Vol. I. University of Connecticut, Storrs. 778 pp.

Slater, J. A., and D. B. Wilcox. 1973. The chinch bugs or Blissinae of South Africa (Hemiptera: Lygaeidae). *Mem. Entomol. Soc. S. Afr.* 12:135.

Vane-Wright, R. I., C. J. Humphries, and P. H. Williams. 1991. What to protect?—systematics and the agony of choice. *Biol. Conserv.* 55:235–254.

Williams, P. H., C. J. Humphries, and K. J. Gaston. 1994. Centres of seed-plant diversity: the family way. *Proc. Roy. Soc. London* (B) 256:67–70.

Zimmerman, Elwood C. 1991–1994. *Australian weevils (Coleoptera: Curculionidae).* 6 Vols. CSIRO Publications and Australian Entomological Society.

Suggested Readings

Gaston, K. J. (ed.). 1996. *Biodiversity: A Biology of Numbers and Difference.* Blackwell Science, Oxford. 396 pp. [A treatment of general biodiversity issues]

GLOSSARY

additive binary coding   a method of coding multistate characters that allows for representation of branching patterns through the use of multiple two-state variables (*see* nonredundant linear coding)

additive character   the form of analysis under which state-to-state ordering is specified; changes hypothesized during cladistic analysis must conform to the original coding or additional steps will be added

ad hoc hypothesis   an assumption invoked to dispose of observations that do not conform to some preferred theory; in cladistics, used in reference to a priori invocations of homoplasy to explain similarity as non-homologous

adjacency   the relative position of character states, one to another, for a multistate character

advanced   used in reference to character data for describing a relative condition, namely as opposed to primitive; the derived condition of a feature; apomorphic

algorithm (for phylogenetic analysis)   a decision-making process for computing cladograms, as in the "Wagner algorithm"

alignment   the process of arranging sequence data from different organisms so that the homologous nucleotide positions correspond to one another across taxa

anagenesis   change within a lineage over evolutionary time, as opposed to cladogenesis, the splitting of lineages

ancestor   *see* hypothetical ancestor

ancestral   *see* primitive

apomorphic   advanced, as opposed to primitive, as an apomorphic character; a character unique to a group and therefore group defining

apomorphy   an advanced character; a group-defining character

area cladogram   in studies of historical biogeography, a cladogram in which the areas where taxa occur are substituted for the taxa themselves

area of endemism   the congruent distributional limits of two or more species (Platnick)

asymmetrical (classification)   a hierarchic scheme of relationships in which branching always occurs in just one of the lineages arising from each successive node (level in the hierarchy) (*see* symmetrical classification)

autapomorphy   a derived character unique to a taxon (*see* apomorphy)

available name   a scientific name that meets certain criteria, specified in the codes of nomenclature, such a having been published in an acceptable manner, having been accompanied by a description of the biological material on which it was based, and others

binary character   *see* two-state character

binomial nomenclature   the system of naming codified by Carolus Linnaeus, in which each species is recognized by a name composed of two words, the generic part (epithet) and the specific part (epithet), as *Homo sapiens*

binominal nomenclature   *see* binomial nomenclature

biodiversity   a term with varied meanings; often used with reference to taxon richness, as numbers of species or higher taxa in a given area

biogeography   the study of geographic relationships among organisms, in the present work used with strict reference to history

bootstrap(ping)   a technique that uses resampling and replication of characters in an attempt to understand to what degree a data set supports a given tree topology

branch-and-bound algorithm   a phylogenetic algorithm that determines an exact result (guaranteed most parsimonious tree) by examining only a portion of the universe of all possible trees, having discarded those portions of that universe that contain only trees longer than any of the trees in the portion examined in detail

branching character   a multistate character in which adjacency relationships in a given direction are multiple (*see* linear character)

branch swapping   a technique used in numerical phylogenetic computations that improves the chances of finding most parsimonious trees; "branches" from potentially useful trees are moved to different locations in an attempt to find shorter trees, the length of the tree being recomputed with each move

center of origin   in biogeography, the area in which a group presumably originated and from which it later spread

character   a feature showing group-defining variation

character state   one of the various conditions of a feature (character) observed across a group of taxa

character-state tree   the graphic representation of the coding of a character; the topological relationships of the states of a character as coded

chorological progression   within a group of taxa, the progressive advancement of characters with increase in distance from the geographic center of origin

cladist   one who practices systematics using the methods of cladistics, i.e., grouping by synapomorphy through the application of the parsimony criterion

cladistics   grouping by synapomorphy through the application of the parsimony criterion

cladogenesis   the spitting of lineages over time, with the consequent increase in numbers of taxa (*see* anagenesis)

cladogram   a depiction of hierarchic relationships in the form of a treelike diagram, with the intent of showing relative recency of relationship, without the connotation of amount of difference

classification (biological)   formalizing the results of phylogenetic analysis, usually by assigning ranks as used in the Linnaean hierarchy; assigning rank, and therefore relationship; sometimes, a cladogram

clique   the set of perfectly congruent characters forming the optimal result in a compatibility analysis; also, maximal clique (*see* compatibility analysis)

clustering-level distance   an ultrametric, or Euclidean, distance of the type employed in phenetics; such a distance implies a uniform rate of divergence by taxa that cluster at the same level (*see* path-length distance)

co-evolution   the proposition that taxa have evolved in concert, as for example hosts and their parasites, with the expectation of congruent patterns of relationships

compatibility analysis   a technique for reconstructing relationships among taxa, whereby all characters contributing to the result of the analysis must be perfectly congruent with the result; that set of characters is called a "clique" or "maximal clique"

component   a node, and the branches descending from it, in a cladogram; *see* hypothetical ancestor; term

component analysis   the analysis of component information within and among cladograms, as first proposed by Nelson and Platnick; a frequently used approach in historical biogeographic analyses

composite coding   the approach of coding character data in a multistate format where possible, as opposed to presence–absence (reductionist) coding

compromise tree   a tree derived from techniques that allow trees to be combined, but the results do not summarize exactly the groupings from all of the input trees (*see* consensus; consensus tree)

concordance   congruence

congruence   the property of two or more characters describing the same hierarchic scheme; sometimes used to mean not in conflict with a given hierarchy

consensus   the collection of groups (components) that is contained exactly in all (most parsimonious) trees resulting from a phylogenetic analysis (*see* compromise tree; consensus tree)

consensus tree   the tree depicting the consensus

consilience   agreement, usually among results from different analyses; referred to by Hennig as "reciprocal illumination"

consistency   the degree to which a character describes accurately a particular hierarchic scheme (*see* consistency index)

consistency index   a measure of consistency, maximal inconsistency having a value approaching 0, perfect consistency having a value of 1; those values are computed as the ratio of minimum possible number of changes in a character on a tree divided by the observed number of changes in the character

cost   the value assigned to a state-to-state change during cladistic analysis

data matrix   information in tabular form on characters for a set of taxa, with the rows representing the taxa and the columns representing the characters

deduction   reasoning from a premise to a logical conclusion, from the general to the specific (*see* induction)

definition   an enumeration of the apomorphies for a taxon (*see* diagnosis)

derived   *see* advanced

description   in systematics, a detailed written statement of the attributes possessed by a given taxon

diagnosis   in systematics, a summary statement of attributes that allows recognition of a taxon and separation of that taxon from one or more other taxa

dichotomy   a node on a cladogram from which two branches arise (*see* polytomy; trichotomy)

discordance   disagreement, as for example the degree to which a multistate character does not agree with a cladogram it helps define

dispersal   in biogeography, the method by which taxa increase the size of their range

distance a measure of similarity (or divergence) among taxa (*see* path-length distance; clustering-level distance)

endemicity *see* endemism

endemism in biogeography, the idea of a taxon (taxa) being restricted to a place

ensemble consistency index the consistency index for the suite of all characters used in computing a cladogram

essentialism a claim about the existence of hidden structures that unite diverse individuals into natural kinds (Sober); the school of thought that assigns innate attributes to taxa without regard to the inherent variability of biological systems (after Mayr)

Euclidean distance *see* clustering-level distance

evolutionary taxonomy school of systematic (taxonomic) practice in which paraphyletic groups are recognized because some of their members show substantial divergence from all other members of the group; a syncretistic approach which combines ideas from both cladistics and phenetics

exact solution (to a phylogenetic problem) solutions provided by algorithms that examine all possible trees for a given data set or that produce equivalent results using branch-and-bound algorithms

exemplar taxon a species, or other lower-level taxon, chosen to represent a higher-level taxon

extrinsic data data derived from sources not subject to genetic inheritance; for example, the host plant associations of phytophagous insects (*see* intrinsic data)

falsificationism *see* deduction

Farris optimization *see* additive character

fit consistency; the degree to which characters conform to (define) a cladogram they have been used to compute; often measured with the consistency index

Fitch optimization *see* nonadditive character

fittest tree the tree(s) computed according to the "implied weights" of the characters, so as to maximize total fit

gap in DNA sequence alignment, space inserted to achieve correspondence of nucleotide positions across a group of taxa that possess unequal numbers of nucleotides

genealogy phylogeny; relationships of ancestry and descent

gene tree terminology from the molecular systematic literature for a scheme of relationships determined by data comprising exclusively amino acid and/or DNA sequence data (*see* species tree)

ground plan the set of attributes (characters) possessed by the hypothetical common ancestor of a group of taxa; may be deduced, as in a composite taxon used as an outgroup, or optimized, as for an HTU (node) on a cladogram computed from actual character data

heterobathmy (of synapomorphy) the phenomenon of different characters defining groups at different levels in the taxonomic hierarchy

heuristic (phylogenetic solution) a numerical cladistic result determined through the use of algorithms applied to data sets so large that all possible trees cannot be examined (*see* exact solution)

holomorph the totality of all characters pertaining to a taxon (Hennig)

holophyletic monophyletic; a term coined by evolutionary taxonomists to allow for the use of monophyletic in reference to paraphyletic groups

holophyly   the property of being holophyletic

holotype   the unique specimen designated to represent the concept for a named species; the name bearer for a taxon of the species group

homology   the recognition across taxa of identity among structures, behaviors, and other attributes, on the basis of similarity of form and position

homonym   the same name applied to two or more groups (e.g., species, genus, family) in botany or zoology, but not between the two fields

homoplasy   incongruence; convergence; parallelism; the presumed multiple origin, reduction, or re-evolution of structures or behaviors

HTU   *see* hypothetical taxonomic unit

hypothetical ancestor   the node in a cladogram representing the deduced set of attributes for two or more terminal taxa

hypothetical taxonomic unit (HTU)   internal node on a cladogram; *see* hypothetical ancestor; node

identification   the process of assigning specimens to names

implied weighting   an optimality criterion whereby weights (values) of characters are determined by the degree to which the characters "fit" the tree, the optimal tree being the one with the greatest sum for the weights of all characters (Goloboff) (*see* fit-test tree)

inapplicable data   data that can never be known for a given taxon and therefore cannot be coded

incertae sedis   literally, "of uncertain position"; applied to taxa, which for some reason, cannot be placed with certainty in a classification, as for example, incomplete fossils

incongruence   lack of agreement, as usually applied to the fit of characters to a cladogram (*see* congruence)

indel   "insertions and deletions," in reference to DNA-RNA sequence data

independence   the quality of data that makes them appropriate for conjoint analysis, whereby in the case of characters, variation in one character is not tied to variation in other characters; the ability of different characters in phylogenetic analysis to serve as unique sources of evidence

induction   reasoning or drawing a conclusion from particular facts or observations; reasoning from the specific to the general; the Baconian method (*see* deduction)

intrinsic data   data subject to genetic inheritance; for example, morphological features or DNA sequences (*see* extrinsic data)

isomorphy   identity; the term used to describe the exact portrayal of relationships in a ranked formal hierarchic classification as found in a cladogram

jackknife (jackknifing)   a scheme of subsampling characters and computing the consensus of all trees from all subsamples, as an attempt to determine the ability of various portions of the data to define a consistent result

lectotype   a specimen, serving the function of a holotype, as designated from the members of a syntype series

lineage   a terminal taxon or monophyletic group

linear character   a multistate character in which adjacency relationships are always singular; an ordinal character (Pimentel and Riggins) (*see* branching character)

Linnaean hierarchy   the hierarchic system of named ranks codified by Linnaeus and elaborated by later workers

long-branch attraction   the theory that terminal taxa may contain large numbers of nucleotide sequences evolving in parallel, and which may therefore group together even though those terminal taxa are not each others' closest relatives

Manhattan distance   path-length distance

mapping (of characters)   the practice of plotting "extrinsic" character data on a cladogram produced from a matrix of "intrinsic" data; more generally, determining the distribution of any character not used in a prior analysis

matrix   *see* data matrix

maximum-likelihood estimation   determination of relationships among organisms through computation of probabilities of character distributions (evolution) on the basis of some predetermined model of character evolution

metric   a measure; in systematics, metrics satisfying the triangle inequality are used as the basis for computing distances in cladistics and phenetics (*see* clustering-level distance; path-length distance)

monophyletic   a group defined by synapomorphies; a group containing a hypothetical common ancestor and all of its descendants (*see* paraphyletic; polyphyletic)

morphocline   character states ordered on the basis of relative similarity alone, without regard for congruence with other characters

mosaic evolution   characters apparently distributed in an unorderly way and not conforming to a hierarchical scheme

most parsimonious tree (MPT)   for a given data set, the tree(s) of minimum length as computed under the parsimony criterion

multistate character   a feature for which there are three or more conditions in a set of three or more taxa

natural group   a monophyletic group

neotype   a specimen selected to serve the function of the primary type, this being required because the holotype, lectotype, or syntypes have been lost or destroyed

nesting   the property of hierarchy, whereby smaller, less inclusive groups, are subsets (i.e., completely included in) of larger, more inclusive, groups

network   an undirected hierarchic arrangement of taxa; an unrooted result of numerical cladistic computation

node (on a cladogram)   the point of intersection in a hierarchy that identifies a component; a hypothetical taxonomic unit

nomenclature   *see* binomial nomenclature

nonadditive character   the form of analysis under which the same cost (length) is assigned to any state-to-state change for a character during cladistic analysis

nonredundant linear coding   a method of character coding that uses multiple variables to represent complex multistate characters, but those variables need not all be two-state, as in additive binary coding; *see* additive binary coding

numerical taxonomy   the name originally attached to phenetics; the theory and practice of grouping by overall similarity with the attendant assumption of uniform rates of change; sometimes, any taxonomic approach that applies quantitative techniques

objective character weighting   weighting on the basis of the observed behavior of characters, using measures such as consistency or fit (*see* subjective character weighting)

ontogeny   development; the stages of the life cycle of an organism

operational taxonomic unit (OTU)   a terminal taxon used in an analysis of relationships, especially as understood in the practice of phenetics

optimality criterion   the standard used to evaluate a result in phylogenetic analysis; for example, the optimality criterion for computing most parsimonious trees is minimization of required character-state changes under a given model of change (*see* additive character; nonadditive character)

optimization   the method of assigning character states to hypothetical taxonomic units (cladogram nodes)

ordered character   *see* additive character

ordinal character   *see* linear character

outgroup   the taxon (or taxa) used to determine the root of a cladogram (or other diagram of relationships among taxa, as for example those produced by maximum-likelihood analysis), and thus the polarities of the characters used to compute it (*see* root)

overall similarity   the concept of relationship based on similarity and difference; the method used for grouping by the phenetic approach, under which the shared absence (as opposed to loss) of characters may be deemed informative

paralogy   nonhomologous duplication; used to describe portions of the genome that exist in multiples and that potentially may be confused with one another; by analogy, in biogeography and co-evolution, the same area occurring more than once in sequence on an area cladogram

paraphyletic   a group containing a hypothetical common ancestor and some, but not all, of its descendants (*see* monophyletic; polyphyletic)

parenthetical notation   a method of describing the nesting of taxa in a hierarchy for use by computer phylogenetics programs

parsimony   simplicity of explanation; minimizing ad hoc hypotheses; the approach applied in cladistics whereby similarities are assumed to be homologous, in the absence of evidence to the contrary

path-length distance   distance along a path between two points; in cladistics, measured in "steps" (numbers of character state changes) without assumptions about rates of divergence among taxa (*see* clustering-level distance)

pectinate (classification)   a cladogram that branches asymmetrically, like the teeth in a comb

pheneticist   one who practices phenetics

phenetics   the method(s) of classifying organisms whereby rank and relationship are determined on the basis of overall similarity (i.e., the sum of similarities and differences) and uniform rates of change are assumed

phenogram   a diagrammatic representation of hierarchic relationships derived from the application of phenetic techniques, whereby taxa are clustered at levels

phylogenetic systematics   cladistics, often more particularly as outlined by Willi Hennig

phylogeny   the genealogical relationships among a set of taxa; sometimes, the process of evolutionary diversification

plesiomorphic   primitive, as opposed to advanced; the quality of being group-defining only at a higher level

plesiomorphy   a primitive character, not group defining at the level at which it is being observed; the quality of being primitive

polarity   the direction of character change

polyphyletic   a group of taxa not including their hypothetical common ancestor (*see* monophyletic; paraphyletic)

polytomy   four or more branches arising from a single node on a cladogram (*see* dichotomy; trichotomy)

primitive   relatively earlier; used in reference to the appearance of characters (or their states) on a cladogram; plesiomorphic

prediction   deduction from a theory; the anticipated nature of a future observation (*see* deduction)

presence–absense coding   an approach to character coding whereby all data are rendered in a two-state format

priority   the precept of biological nomenclature stipulating that the name first applied to a taxon is the one that will be treated as valid

ranking   the assignment of hierarchical position

reciprocal illumination   consilience

reductionist coding   the practice of coding all character data in a presence–absence format (*see* composite coding)

retention index   the fraction of potential synapomorphy retained as synapomorphy on a cladogram

root   the point at which a cladogram is given direction; the taxon, or outgroup, used to determine the polarity of the characters used to compute a cladogram (*see* outgroup)

rooting   determining the root; the position of the root

scattering   the observed behavior of character-state distributions when a two-state character does not fit a cladogram perfectly

scenario   a story describing the evolution of organisms, structures, etc.; usually not based on critical analysis of evidence

sequencing (in classifications)   the treatment of successive sister-group relationships as being of equal rank; a method for minimizing the number of ranks required to convert a cladogram into a formal classification (*see* subordination)

sequencing (of nucleotides)   the determination of nucleotide composition and order in a portion of the genome

sister group(s)   a pair of taxa united by one or more unique characters

special similarity   *see* synapomorphy

species   in cladistics, often used in reference to the minimal-level taxon subject to analysis

species tree   terminology from the molecular systematic literature for a phylogenetic scheme based on data other than DNA and amino acid sequences (*see* gene tree)

step(s)   on a cladogram, the number of state changes for a character or characters; the measure of length used when computing a path-length (Manhattan) distance

subjective character weighting   determining the value of characters for phylogenetic analysis on the basis of criteria not related to the behavior of the characters themselves, as for example, apparent morphological complexity or presumed adaptive significance

subordination (in classifications)   the treatment of successive sister-group relationships as being of progressively lower rank (i.e., the nesting of ranks corresponding the nesting of the hierarchy itself); the method requires the maximum number of ranks to convert a pectinate cladogram into a formal classification (*see* sequencing in classifications)

subtree analysis   an approach in historical biogeography for analyzing congruence among area cladograms, whereby paralogous areas are removed before comparisons among the area cladograms are made

successive approximations weighting   a method of evaluating the strength of characters as they define a given cladogram(s); a most parsimonious tree(s) is first computed, then characters are weighted, often according to their consistency index on that tree[s], and the tree recomputed; the weighting and recomputation process is repeated until a stable result it achieved (*see* implied weighting)

surrogacy   in some studies of biodiversity, the use of one or a few taxa as a way of assessing the characteristics of a biota without the necessity of sampling all or most of its constituent members; selective sampling

symmetrical (classification)   a branching pattern in which subdivision occurs equally in all lineages at successive nodes (*see* asymmetrical classification)

symplesiomorphy   shared, primitive, traits defining groups only at higher levels

synapomorphy   shared, derived, group-defining trait

syncretist   an evolutionary taxonomist; one who combines parts of different methods (e.g., phenetics and cladistics)

synonym   two or more different names applied to the same taxon

syntype   two or more specimens examined by the original author of a species, none of which was uniquely designated to serve as the name bearer for the taxon

systematics   the practice of recognizing taxa, determining hierarchic relationships among those taxa, and formally specifying those relationships; frequently used in a sense roughly equivalent to taxonomy

taxon (taxa)   a grouping of organisms at any level in the systematic hierarchy

taxon-area cladogram   *see* area cladogram

taxonomy   the practice of recognizing and classifying organisms; frequently used in a sense equivalent to systematics

term   a terminal taxon; in biogeography, a terminal area in an area cladogram (*see* component)

terminal taxon   a taxon for which data are actually coded in a cladistic analysis; a group of organisms that for the purposes of a given study is assumed to be homogeneous with respect to other such groups

test   comparison of observation with theory, as comparing the actual distribution of characters with a scheme of phylogenetic relationships (*see* deduction)

three-taxon statements   a method of coding whereby information on taxa is coded to represent the number of three-taxon statements the data for those taxa imply

tokogenetic relationships   the reticulate pattern of connections existing below the species level as a result of interbreeding or other factors

topology (of a cladogram)   the geometric form of a cladogram; the pattern of branching of a cladogram

transformation series   a specified ordering of relations between the states of a character

tree   generally, any branching diagram that specifies hierarchic relationships among taxa; sometimes, a branching diagram specifying ancestor–descendant relationships or patterns of speciation (after Nelson)

triangle inequality   the property of triangles stipulating that the sum of the lengths of two sides cannot be less that the length of the third side

trichotomy   three branches arising from a single node on a cladogram (*see* dichotomy; polytomy)

two-state character   a character for which there are two conditions

type specimen   *see* holotype

ultrametric   *see* clustering-level distance
unordered character   *see* nonadditive character
valid name   the available name correctly applied to a taxon
variable   a character; the representation of a character (and its states) in a data matrix
vicariance   in biogeography, the subdivision of ancestral ranges of taxa
Wagner algorithm   a method of computing phylogenetic relationships using the parsimony criterion; the underlying basis of all parsimony-based phylogenetic algorithms
weighting (of characters)   assigning relative importance to characters on the basis of some criterion (*see* objective character weighting; subjective character weighting)

SUBJECT INDEX

a posteriori character weighting, 129
a priori character weighting, 128
abstracting sources, 24
Adams compromise, in co-evolution, 195
Adams (compromise) technique, 147
Adanson, Michel, 6
adaptation, testing hypotheses of, 204, 207
additive binary coding, 97, 100, 130
additive transformation, 93, 117
additivity, 93, 117
adjacency. *See* character adjacencies
advanced, 79
advanced characters, 206
age, minimum, 84
age, relative, 84
algorithms, phylogenetic, 125
alignment
    computer implemented, 104
    of nucleotide sequences, 73, 105
    parsimony based, 104
    similarity based, 104
    as statistical problem, 74
    using models, 104
allozyme data, 105
Amniota, classification of, 12
anagenesis, 169
ancestor-descendent relationships, in classifi-
    cations, 171
ancestors
    hypothetical, 84
    paraphyly of, 84
    as specimens, 85
ancestral polymorphism, 156
apomorphic characters, distribution of, 68
apomorphy, 12, 65
approximate solution algorithms. *See* heuris-
    tic algorithms
area cladograms
    construction of, 189
    discussion of, 186, 187, 189, 191
    resolved, 189, 191

area paralogy, 192
areas of endemism
    discussion of, 186, 187
    in Northern Hemisphere, 212
    size, 211
Aristotle, 3
Assumption 0, in biogeography, 190
Assumption 1, in biogeography, 190, 191
Assumption 2, in biogeography, 190, 191
assumptions, ad hoc, 48
Asteraceae, 200
asymmetrical branching, of trees, 150
autapomorphic characters, 12, 65, 99
autapomorphies, in data matrix, 100
authorship, of names, 30
availability, of names, 33

Bacon, Francis, 46
behavioral characters, 201
binary variables, 98
biodiversity, 209
biogenetic law, discussion of, 81
biogenetic law, validity of, 82
biogeographic assumptions, 189
biogeographic connections, intercontinental,
    184
biogeographic regions, 182
biogeographic relationships, 211
biogeography
    cladistic, 184
    early history, 179
    historical, 179
Biological Abstracts, 25
biology
    comparative, 18
    divisions of, 18
    general, 19
biosystematics, 15
biotic regions, of Sclater, 182
bird pollination, origins in Mirbelieae, 205
Blissidae, of North America, 212

Blissidae, of South Africa, 212
bootstrapping, 160
boreal megafaunal bias, 211
branch and bound algorithms, 125
branch lengths, in maximum likelihood, 138
branch swapping algorithms, 127
branching character format, 97
Bremer support, 157
Bremer support values, 158
Brooks Parsimony Analysis, 190
Brundin, Lars, 11
Bulletin of Zoological Nomenclature, 42

Caesalpino, Andrea, 3
catalogs, systematic, 21, 23
center of origin, 186
character adjacencies, 82
character analysis, 89
character, concept of, 89
character polarity
    chorological progression, 79
    common equals primitive, 79
    determination of, 79
    direct method, 79
    fossils are primitive, 79
    indirect method, 79
    simple is primitive, 79
character weighting
    compatibility based, 130
    consistency based, 130
    objective, 129
    subjective, 128
    successive approximations, 130, 150
character-coding methods, assumptions of,
    95
characters
    discussion of, 20
    rechecking and correcting, 49
    self-consistency of, 131
    value of complex, 49
    value of simple, 49
checklists, 21
Chironomidae, 212
Chironomidae, distribution of, 185
Chrysomelidae, 200
cladist, 8
cladistic biogeography, criticisms of, 193
cladistic classifications, arguments against,
    166
cladistics, 8, 24
cladogram, 8, 10, 84, 85
cladograms, comparison of, 203
classification(s)
    definition of, 15
    formal, 165
    hierarchic, 17
    with numbered ranks, 167

pectinate, 169
    transmission of information by, 169
    unifying principle of, 57
clique analysis. *See* compatibility analysis
cliques, maximal, 114
clustering-level distance, 120
codes of nomenclature, 31
coding, of characters, 90
co-evolution, 179
collections, systematic, 25
colonization, of hosts, 201
combinable components, 147
combined analysis. *See* total evidence
    analysis
common plan, 64
compatibility, 114
compatability analysis, 111, 114
compatability of characters, weighting by, 130
component, 187
components, in biogeography, 192
composite coding, of characters, 95
compromise techniques, 147, 148
compromise trees, 147
concave weighting function, 132
concordance, 112, 114
concordance, of characters, 91, 92
congruence, 70, 114
congruence, of characters, 51
conjunction, 66
connection, by intermediate forms, 69
consensus
    in co-evolution, 195
    discussion of, 146
    evaluation of, 154
    method of analysis, 151–153
consensus technique, 147, 148
consensus tree, 147
conservation, 212
consilience, 114
consistency, of characters, 50
consistency index, calculation of, 112, 113
consistency of characters, weighting based
    on, 130
constant of concavity, 132
continental drift, 180
convergent evolution, 72
corroboration, degree of, 48
corroboration, of theories, 47
co-speciation, 102
cost, of character state change, 116, 121, 122
Croizat, Leon, 182
Curculionidae, of Australia, 212
Curculionidae, of North America, 212

Darlington, Philip J., 181
Darwin, Charles, 4, 15, 21, 180
data decisiveness, 151

data, hierarchic description of, 167
data matrix, 96
data sets, very large, 133
database, Biota structure, 176
databases, systematic, 174
deductive approach, 46
definition, concept of, 56
definition, of groups, 56
deletions, in nucleotide sequences, 103
description, of new taxa, 22
diagnoses, exceptions to, 165
diagnoses, succinctness, 165
diagnosis, of a taxon, 22
dichotomous branching in cladograms, 50
discordance, of character, 92
dispersal, ad hoc theories of, 186
dispersal, in biogeography, 180
distance data, 106
distance levels, 122
distances, observed, 122
distributions, disjunct Southern Hemisphere,
    185
diversity, measure of, 214
DNA hybridization, 106
du Toit, Alexander, 181

empirical testing, of a theory, 47
endemicity, 213
endemism, determining areas of, 188
ensemble consistency index, 112
enumeration of biotas, 209
epithets, 30
essence, Aristotelean, 56
essentialism, 41, 56
Euclidean distance, 120
evolution, 45
evolutionary scenarios, 199
evolutionary taxonomy, 4
evolutionary tree, 10
exact solution algorithms, 125
exemplar method, of taxon representation,
    107
exemplar terminals, 101
exhaustive search algorithms, 125
explanation, 58
explanatory power, 55
extrinsic data, 102

fact, 44
Fahrenholz's Rule, 194
falsifiability, 47
falsification, of theories, 47
falsificationism, absolute, 51
falsificationism, naive, 51
falsifiers, of ontogeny, 82
family-group names, 32
family-group types, 35

Farris, J. S., 122
Farris optimization, 93, 117
Farris transformation, of character states,
    93, 117
fit, 99, 112, 116, 131
Fitch optimization, 93, 117
Fitch transformation, of character states, 92,
    93, 117
fittest trees, 131, 151
fixation, of genus-group type, 35
fixation of type, by monotypy, 35
fixation of type, by subsequent designation,
    35
flora, of North America, 211
flora, of South Africa, 211
flora, of Western Australia, 211
fossil taxa, classification of, 171
fossils, as ancestors, 84
fossils, intermediate, 63
frequency data, 105
fusion of terminals, 101

gaps, in alignments, 104
gender agreement, in nomenclature, 40
gene trees, 156
genealogical relationships, 210
general area cladogram, 186, 187
general area cladograms, preparation of, 192
general (compromise) technique, 147
genus-group names, 32
genus-group types, 35
Geoffroy St. Hilaire, Etienne, 64
geology, influence on biogeography, 180
Geomyidae, 195
Gerridae, 200, 202
gradist, 8
groundplan, deduced, 108
groundplan method, of taxon representation,
    107
groundplan, optimized, 108
groups, unnatural, 69
Gymnospermae, as paraphyletic group, 14

Hennig, Willi, 11
heterobathmy of synapomorphy, 69
heuristic algorithms, 127
hierarchical discordance, of a character, 91
hierarchical pattern, 156
hierarchy, information in, 165
historical linguistics, 8
holophyly, 76
holotype, 34
homogeny, 71
homologous features, distribution of, 65
homology
    concept of, 63, 71
    criteria, 64

homology (*continued*)
  definition of, 64
  primary, 71
  tests of, 66
  usages of, 71
homology analysis, 93
homonyms, replacement of, 36
homonymy, in nomenclature, 35, 67
homoplasy, 65, 71, 72, 114
Hooker, Joseph Dalton, 180
horizontal gene transfer, 156
host associations, analysis of, 196, 201
host switching, 200
host-parasite co-evolution, 194
HTU. *See* hypothetical taxonomic unit
hypotheses, 44
hypotheses, ad hoc, 48
hypothetical ancestor, 85
hypothetical taxonomic unit, 84

identification, definition of, 18
identification keys, 16
implied weighting, 131
inapplicable data, 99
incertae sedis taxa, in classifications, 171
inclusive groups, recognition of, 20
incongruence, 82, 156
incongruence, in biogeography, 194
inconsistency, under parsimony, 54, 139
indels, 103
independence, of characters, 98
independence, of data, 154
Index Herbariorum, 25
Index Kewensis, 21, 42, 175
indexing sources, 24
induction, problem of, 47
inductive approach, 46
information content, of classifications, 149
inheritance, mechanisms of, 63
Insecta, classification of, 165, 168
insects, holometabolous, 141
insertions, in nucleotide sequences, 103
internal rooting, of multistate characters, 97
International Botanical Congress, 31
International Commission of Zoological
    Nomenclature, 31
intrinsic data, 102
introgression, 156
intuitive method, of taxon representation,
    107
islands, of trees, 133
isomorphy, of classifications, 165
isotype, 35

Jac program, 134
jackknife resampling, 134
jackknife results, 159

jackknifing, 157
junior homonym, 36
junior synonym, 37

keys, 17
keys, dichotomous, 16
knowledge, objective, 44

land bridges, 180
Latin, 29
lectotype, 34
legumes, 204
levels of grouping, 69
linear character format, 96
Linnaean hierarchy, criticisms of, 40
Linnaean hierarchy, discussion of, 38, 39
Linnaeus, Carolus, 3, 209
Linsley, Gorton, 5
literature, systematic, 20
literature, theoretical, 24
local optimum, 133
logical consistency, of a theory, 47
logical form, of a theory, 47
long-branch attraction, 139
loss, as synapomorphy, 70

majority rule (compromise) technique, 147
Manhattan distance, 120
mapping, of attributes onto cladograms, 200,
    202
matrix. *See* data matrix
Matthew, William Diller, 181
maximum likelihood
  calculation of, 138
  discussion of, 55, 136
  inapplicability to morphology, 139
Mayr, Ernst, 5, 12
measurement data, 107
measures of support, 157
metric, 106
metricity, 120
Mirbelieae, 204
Miridae, Australian diversity, 210
Miridae, North American diversity, 210
missing areas, in biogeography, 190
missing data, 99
missing taxon, 187
models of evolution, under maximum likeli-
    hood, 138
monographs, 21
monophyletic group(s)
  in classifications, 172
  definition of, 74
  depiction of, 75
  under Farris' definition, 75, 76
monophyly, 76
morphocline analysis, 92

morphometrics, 107
mosaic evolution, 69
most parsimonious trees, 127, 150
multiple cladograms, 146
multistate characters, coding of, 91, 94, 95

named groups, in classifications, 172
natural groups, 66
neighbor joining, 134
nelsen command, 147
Nelson, Gareth, 11
Nemcia, 204
neotype, 34
nest architecture, in Polistinae, 203
nesting, of synapomorphies, 149
networks, unrooted, 77, 78, 122, 136
new combination, in nomenclature, 40
node, in a cladogram, 97, 98, 124
Nomenclator Zoologicus, 42
nomenclature
    aids to, 42
    binomial, 29
    codes of, 31
    stability in, 31
nonadditive character transformation, 92, 93,
    117
nonadditivity, 117
nonmetricity of data, 106
non-protein-coding gene regions, 103
nonredundant linear coding, 97–100
nucleotide sequences, homology in, 73
numerical taxonomy, 6, 8

objective character weighting. *See* a posteri-
    ori character weighting
objective synonym, 37
Occam, William of, 48
ontogenetic data, example of, 80
operational taxonomic unit, 19
Ophraella, 200
optimal trees, 122
optimality criteria, 116
optimization
    of additive characters, 118
    in co-evolution, 204
    concept of, 116, 201
    of non-additive characters, 119
optimization alignment of DNA sequences,
    104
order, of character states, 77
ordered. *See* additivity
ordered transformation, 93
ordinal subset, of character states, 98
orthologous genes, 74
OTU. *See* operational taxonomic unit
outgroup, 77
outgroup comparison, 77

outgroups, choice of, 83
Owen, Richard, 63

paleontological approach, 84
Papilionidae, 200
Papilionidae, host associations of, 201
Papilionidae, phylogenetic relationships of,
    201
parallel evolution, 72
parallelism, usages of, 71
paralogous distributions, in biogeography,
    187, 193
paralogous gene families, 156
paralogous genes, 74
paraphyletic group(s)
    definition of, 75
    depiction of, 75
    under Farris definition, 75, 76
parasitological method, 102
paratype, 35
parenthetical notation, 122, 123
parsimony
    with additive characters, 117, 122
    Dollo, 116, 122
    first use in systematics, 116
    generalized, 117
    with irreversible change, 116, 122
    as methodological criterion, 48, 53
    with nonadditive characters, 117, 122
    presuppositions of, 53
    with Sankoff characters, 117, 122
parsimony ratchet, 135
path length distance, 120
pattern cladist, 58
patterns, in systematics, 58
pectinate classification, 169, 171
permutation tail probability, 160
pheneticist, 8
phenetics, 6, 8
phenogram, 8, 10
philosophy of science, 44
Phthiraptera, 195
Phyllobrotica, 195
phylogenesis, 169
phylogenetic systematics, 11
phylogenetic taxonomy, 41, 173
phylogenetic uniqueness, 213
phylogenetics, 7
phytophagous insects, 200
Pilophorini, 212
plesiomorphic characters, distribution of, 68
plesiomorphy, 65
pocket gophers, parasites of, 196
pocket gophers, phylogenetic relationships,
    196
polarity, of character states, 77
Polistinae, 201

Polistinae, phylogenetic relationships of, 203
pollination, by birds, 204
polymorphisms, coding, 101
polyphyletic group(s)
    definition of, 75
    depiction of, 75
    under Farris definition, 75, 76
polytomy, 58
Popper, Karl, 46
Popperian approach, criticisms of, 51
positivism, 46, 51, 90
prediction, in classifications, 165
prediction, in systematics, 52
presence-absence coding, 95
primary homonym, 36
priority, in nomenclature, 32
problem solving in science, 46
progression rule, 79, 183, 186
Proteaceae, 183, 212
Proteaceae, biogeographic connections, 184
Proteaceae, biogeographic tracks in, 183
protection of radiating lineages, 213
protection of the most lineages, 213
protein-coding gene regions, 103
PTP. *See* permutation tail probability
publication, criteria of, 33

randomization tests, 160
rank, in classifications, 167
ratchet. *See* parsimony ratchet
rate of change, under parsimony, 55
rates, evolutionary, 8, 54
rates, of nucleotide substitution, 140
ratios, as characters, 107
Ray, John, 3
reductionist coding, of characters, 94
redundant distributions, in biogeography, 187
relational database, systematic model, 175
Reptilia, 13
rescaled consistency index, 112, 113
retention index, 114, 115
revision, taxonomic, 21, 22
rooting, of trees, 77, 135

scattering, of a character, 90, 91
scenarios, 53
scenarios, methods of testing, 199
schools, of systematics, 8, 9
scientific names, categories of, 33
Sclater, biogeographic regions of, 182
secondary homonym, 36
self consistency, of characters, 131
senior homonym, 36
senior synonym, 37
sequencing convention, in classifications,
    170–172
serial homology, 67

similarity, of structures, 69
similarity, special, 7, 12
Simpson, George Gaylord, 181
simultaneous analysis. *See* total evidence
    analysis
singular statement, 47
sister group(s)
    age of, 84
    concept of, 84
    depicted, 85
Sneath, Peter, 6
social behavior in wasps, 204
Sokal, Robert, 6
southern continents, biogeographic connec-
    tions of, 212
species
    concepts, 19
    diagnosis of, 20
    recognition of, 20
    total number of, 209
Species Plantarum, 29
species trees, 156
species-group names, 32
species-group types, 34
specimens, selection of, 107
Spermatophyta, 78
Spermatophyta, classification of, 13
spider web evolution, 199
stability, in nomenclature, 31
states, of characters, 90
statistical approach, to phylogenetics, 54
stemmatics, 7
steps, on a cladogram, 112, 116, 117
stepwise decisions approach, to character
    coding, 94
storing trees, 122
Strepsiptera, 140
Stricklandian Code, of nomenclature, 31
strict consensus, 147
subjective character weighting. *See* a priori
    character weighting
subjective synonym, 37
subordinated classification, 170
subordination, 171
subtree analysis, 192, 193
subtree pruning-regrafting algorithms, 127
successive approximations weighting, 149
successive approximations weighting, de-
    scription of, 130
suprafamilial names, 31
surrogacy, 213
swamping, 155
symmetrical branching, of trees, 150
symmetry, in classifications, 167
synapomorphy
    concept of, 65
    measure of, 114

nesting of, 66
tests of, 69
usages of, 71
syncretist, 8
synonymy, in nomenclature, 37
syntype, 34
Systema Naturae, 29, 39
Systematic Biology, 24
Systematic Botany, 24
Systematic Zoology, 11
systematics, definition of, 15
systematics, independence of, 58

T-PTP. *See* topology dependent permutation
    tail probability
taxa, selection of, 107
Taxon, 24
taxon, definition of, 19
taxon, terminal, 19
taxon-addition sequence, randomizing, 126
taxon-area cladogram. *See* area cladogram
taxonomic congruence, 152, 153
taxonomy, as art, 5
taxonomy, definition of, 15
term, 187
terminal taxon, 19, 85, 100
terminology, 15
test, of a theory, 47
Tetrapoda, 67, 78
textual criticism, 7
theories, explanatory power of, 52
theories, testing of systematic, 52
theory, 44
theory, causal, 57
Therophosidae, 23
Theropoda, 45
three-area statements, 187, 192
three-taxon method, 136, 137
tokogenetic, 156
topology, of cladograms, 150
topology dependent permutation tail proba-
    bility, 160
total evidence analysis, 151–153
total fit, 131
track analysis, in biogeography, 182
traditional rankings, in classifications, 172
transformation series analysis, 92
transformation series, of character states, 77

transformation, nesting of, 67
transformation, ontogenetic, 63
tree, 84, 85
tree bisection-reconnection algorithm, 127
tree island, 133
tree, unrooted, 136
trees, numbers of rooted bifurcating, 124
triangle inequality, 120
trichobothria, in Reduviidae, 80, 81
trichotomy, 50
true phylogeny, 156
truth, 44
truth, in science, 48
tuatara, 213
two-state character, 91, 95, 113
type concepts, 34
type genus, 37
type species, 35, 37
type specimen, 41

ultrametric, 120, 167
unambiguous branch support, 149
uninomial nomenclature, 41
universal statements, in biology, 47
universal statements, in logic, 47
unordered. *See* nonadditivity
unordered transformation, 93
unrooted trees. *See* networks, unrooted
Usinger, Robert, 5

variables, characters as, 97
Vespidae, 206
vicariance, 183, 186

Wagner algorithm, 124, 126, 127
Wagner ground plan method, 117
Wallace, Alfred Russell, 4, 15, 180
waratahs, biogeographic connections, 183,
    184
Wegener, Alfred, 180
weighting functions, 132
weighting of characters. *See* character
    weighting
widespread taxon, in biogeography, 187, 190
wing polymorphism, 200, 202
Wygodzinsky, Pedro, 11

Zoological Record, 24

AUTHOR INDEX

Acocks, J. P. H., 211
Adams, E. N., 146, 147
Andersen, N. M., 200, 202
Archie, J. W., 114, 160
Ashlock, P. D., 76, 169, 183
Axelius, B., 188

Barbadilla, A. et al., 105
Beatty, J., 57
Bock, W., 6
Boyden, A., 64
Brady, R. H., 55, 58, 64
Bremer, K., 147, 157
Bremer, K. and H.-E. Wanntorp, 11
Brooks, D. R., 102, 195
Brower, A. V. Z. and V. Schawaroch, 74, 90
Brower, A. V. Z. et al., 156
Brundin, L. Z., 79, 181, 183, 192, 212

Camin, J. H. and R. R. Sokal, 116
Carmean, D. and B. Crespi, 140
Carpenter, J. M., 121, 149, 160, 204, 206
Cassis, G. and G. F. Gross, 210
China, W. E., 73
Chippindale, P. T. and J. J. Wiens, 155
Coddington, J. A., 199, 205, 207
Colwell, R. K., 174
Corrick, M. G. et al., 211
Cranston, Peter, 160
Craw, R. C., 183, 193
Crisp, M. D., 199, 204, 205
Croizat, L., 182
Crowson, R. L., 18

Darlington, P. J., Jr., 181, 211
Darwin, C., 70
Davis, J. I, 157
Davis, J. I and K. C. Nixon, 55
Day, W. H. E. et al., 122
de Pinna, M., 72, 74, 156

de Queiroz, A. et al., 154
de Queiroz, K. and J. Gauthier, 56, 173
Dingus, L. and T. Rowe, 45
Dominguez, E. and Q. D. Wheeler, 174
Doyle, J. A. and M. J. Donoghue, 101
Doyle, J. J., 156

Edwards, A. W. F. and L. L. Cavalli-Sforza,
    116
Endler, J. A., 180
Engelmann, G. F. and E. O. Wiley, 53
Enghoff, H., 212
Ereshefsky, M., 19
Eschmeyer, W. N., 21
Estabrook, G. F. et al., 114

Faith, D. P., 160
Farrell, B. and C. Mitter, 195
Farris, J. S., 9, 10, 49, 54, 106, 112, 116, 117,
    121, 122, 150, 160, 165, 171
Farris, J. S. and A. G. Kluge, 114
Farris, J. S. et al., 97, 134, 136
Felsenstein, J., 54, 136, 139, 160
Fichman, M., 180
Fieber, F. X., 72
Fitch, W. M., 117, 156
Futuyma, D. J., 19
Futuyma, D. J. and S. S. McCafferty, 200

Gaffney, E. S., 45, 53, 167
Geoffroy, E., 64
Ghiselin, M. T., 56
Gladstein, D. G., 127
Goloboff, P. A., 125, 127, 131, 151
Goodman, M. et al., 156
Gould, S. J., 106
Guttmann, W., 128

Haeckel, E., 76
Hafner, M. S. and S. A. Nadler, 195

Harvey, P. H. and M. D. Pagel, 200
Hennig, W., 7, 49, 50, 64, 167, 183, 194
Hillis, D. M. et al., 103
Holmgren et al., 25
Huelsenbeck, J. J., 140
Hull, D. L., 6, 51, 55
Humphries, C. J. and L. R. Parenti, 184, 190, 191

Kluge, A. G., 151
Kluge, A. G. and A. J. Wolf, 102, 151
Kluge, A. G. and J. S. Farris, 112, 117, 122
Knox, E. B., 169

Lankester, E., 72
Lanyon, S. M., 157
Le Quesne, W. J., 114, 130
Linnaeus, C., 29, 39
Lyons-Weiler, J. and G. A. Hoelzer, 142

Maddison, D., 133
Maddison, W. P., 156
Margush, T. and F. R. McMorris, 147
May, R., 213
Mayden, R. L., 192
Mayr, E., 5, 56, 64, 166
Mayr, E. et al., 5
McKenna, M. C. and S. K. Bell, 38, 175, 211
Michener, C. D., 41
Mickevich, M. F., 152, 155
Mickevich, M. F. and D. L. Lipscomb, 112
Mickevich, M. F. and C. Mitter, 106
Mickevich, M. F. and N. I. Platnick, 150
Miller, J. S., 200, 201
Mindell, D. P., 74
Mitchell, P. C., 7
Mitter, C. et al., 200
Miyamoto, M. J., 147
Miyamoto, M. J. and W. M. Fitch, 154
Morrone, J. J. and J. M. Carpenter, 192

Neave, S. A., 42
Nelson, G. J., 79, 147, 170, 171, 179
Nelson, G. J. and N. I. Platnick, 56, 136, 179, 184, 190
Nelson, G. J. and P. Y. Ladiges, 192, 193
Nixon, K. C., 135
Nixon, K. C. and J. I Davis, 101
Nixon, K. C. and J. M. Carpenter, 77, 101, 147, 152

O'Grady, R. T. and G. B. Deets, 97
O'Grady, R. T. et al., 97
Owen, R., 63

Page, R. D. M., 192, 195, 196
Patterson, C., 58, 73

Patterson, C. and D. E. Rosen, 172
Pimentel, R. A. and R. Riggins, 94, 97
Platnick, N. I., 23, 57, 79, 90, 169, 184, 187, 211
Platnick, N. I. and H. D. Cameron, 7
Platnick, N. I. and G. J. Nelson, 184, 194
Platnick, N. I. et al., 149
Pleijel, F., 94
Polhemus, D. A., 193
Popper, K., 46, 48

Rieppel, O. C., 46, 64
Rosen, D. E., 81, 184
Ross, H. H., 15

Saitou, N. and M. Nei, 134
Sankoff, D. and R. Cedergren, 117
Sankoff, D. and P. Rousseau, 117
Schuh, R. T., 212
Schuh, R. T. and J. S. Farris, 147, 152
Schuh, R. T. and J. T. Polhemus, 147, 152
Schuh, R. T. and G. M. Stonedahl, 189, 192
Sclater, P. L., 182
Sharkey, M. J., 130
Sherborne, C. D., 42
Sibley, C. G. and J. E. Ahlquist, 211
Sibley, C. G. and B. L. Monroe, 21
Siddall, M. E., 157
Siddall, M. E. and A. G. Kluge, 139
Simpson, G. G., 5, 15
Slater, J. A. and D. B. Wilcox, 212
Sober, E. R., 53, 137
Sokal, R. R. and F. J. Rohlf, 147
Sokal, R. R. and P. H. A. Sneath, 6
Soltis, D. E. et al., 134, 154
Sorensen, J. T. et al., 129
Standley, P. C., 17
Stonedahl, G. M., 16, 22
Swofford, D. L. and W. P. Maddison, 117
Swofford, D. L. et al., 53, 125, 127, 134, 137

Tajima, F., 156

Vrana, P. and W. C. Wheeler, 156

Wallace, A. R., 180, 182
Wegener, A., 180
Wenzel, J. W., 129, 139, 201, 203
Weston, P. H., 81
Weston, P. H. and M. D. Crisp, 183, 184
Wheeler, Q. D., 82
Wheeler, W. C. and D. G. Gladstein, 104
Wheeler, W. C. et al., 154
Whiting, M. F., 142
Whiting, M. F. and W. C. Wheeler, 140
Whiting, M. F. et al., 140

Wiley, E. O., 171, 172, 190
Wilkinson, M., 94
Williams, P. H. et al., 214
Wilson, D. E. and D. M. Reeder, 21
Wygodzinsky, P. and S. Lodhi, 80

Yeates, D. K., 107

Zandee, M. and M. C. Roos, 190
Zimmerman, E. C., 212
Zimmermann, W., 7

SELECTING AND ACQUIRING SOFTWARE

Computer assisted phylogenetic analysis can consist of as many as four distinct activities. These are:

- data matrix preparation
- nucleotide sequence alignment
- phylogenetic computation
- cladogram analysis and printing

Success in performing these activities will depend on your ability to acquire and master suitable software. The factors likely to affect your choice among available options will include: operating system choice, speed, ease of use, and price.

The following paragraphs describe software packages that will perform phylogenetic calculations under the parsimony criterion and which will serve effectively for beginners as well as the most advanced users. Information on additional software packages of lesser or more restricted capability is available on the website of the Willi Hennig Society (www.cladistics.org).

Phylogenetic Computation Software

NONA

This PC-based program is the most powerful phylogenetics package available. It includes a range of options for finding most-parsimonious trees and for evaluating character support. It can be set to address all available RAM and can perform computations on matrices of any size. NONA is the program of choice when performing computations on matrices with more than 128 taxa and 900 characters, which includes many sequence-based data sets, because it can produce results in a sensible time period on standard PC equipment. Its speed and range of options will also save time and facilitate processing of many smaller matrices, because computational time is determined not only by the size of the matrix but also the amount of homoplasy. NONA utilizes similar command and data file structures to Hennig86 (see below). NONA is best used in concert with WINCLADA. (Available at: www.cladistics.com)

PEE-WEE

This program uses the same computational engine as NONA and offers most of the same features. Rather than computing most-parsimonious trees, PEE-WEE produces "fittest trees" under Goloboff's system of implied weights, as described in Chapter 6. (Available at: www.cladistics.com)

PHAST, SPA

These packages allow for computation using "Sankoff characters," whereby any between-state transformation can be assigned a given cost on a differential basis. This approach is most frequently applied in the analysis of nucleotide sequence data. The computational engine is the same as that used in NONA and PEE-WEE and as implemented here allows for effective "combined analysis" because any character can be optimized under any approach. (Available at: www.cladistics.com)

Hennig86

This is an MS-DOS package known for its speed and efficiency. The program is limited to addressing 512k bytes of memory and cannot process more than 128 taxa and 900 characters. It lacks the capability of interpreting nucleotide data coded as nucleotides. For these reasons Hennig86 will be found most effective when dealing with smaller morphology-based data matrices. Hennig86 is capable of finding exact solutions for a larger number of taxa than any other available program. (Information available at: www.cladistics.org)

PAUP

This program package combines phylogenetic computation and data matrix manipulation. It contains a vast selection of options, including the ability to do maximum-likelihood computations, but is significantly slower than NONA in producing results. PAUP, as most users know it, was developed to run under the Macintosh operating system. A beta version, PAUP*, is available for testing, and is said to function on almost any platform. PAUP uses the NEXUS data file format, but can import data in the Hennig86/NONA file format. (Information on availability and purchase at: www.sinauer.com)

Data Matrix Preparation and Cladogram Analysis and Printing

WINCLADA

This program combines data matrix preparation and tree evaluation and printing in a single package; it also allows the user to run the phylogenetics computing software packages NONA, PEE-WEE, and Hennig86 as subprocesses.

WINCLADA provides a versatile data entry/editor which gives the user substantial control over the form in which the data is submitted for calculation of cladograms. It will run in Windows 95, 98, and NT environments. Trees can be viewed and printed in various forms, topologies manipulated, and characters can be displayed in a variety of formats, and optimized under various models. WINCLADA natively reads data in the Hennig86/NONA format, but can read NEXUS files. (Available at: www.cladistics.com)

MACCLADE

This Macintosh program performs tree manipulation and printing in a manner similar to that found in WINCLADA. Consequently, it is often used in concert with PAUP, but can also import trees and data files generated by NONA and Hennig86. (Available at: www.sinauer.com)

Nucleotide Sequence Alignment

MALIGN

This program aligns nucleotide sequence data (and proteins as nucleotides) using parsimony. The program attempts to generate multiple sequence alignments which will yield parsimonious cladograms. There are various search options and output formats consistent with the input of Hennig86/Nona/Winclada and PAUP/MACCLADE. Precompiled PC and Unix executables, documentation, and all source code are available via anonymous ftp from *ftp.amnh.org*/pub/molecular/malign.

POY

This program analyzes both nucleotide and "standard" phylogenetic data directly, searching for parsimonious cladograms without multiple sequence alignment. Nucleotide data are optimized via several direct means (without alignment) and can be combined with standard additive and non-additive character analysis. "Implied" multiple alignments can also be generated. Precompiled PC and Unix executable, documentation, and all source code are available via anonymous ftp from *ftp.amnh.org*/pub/molecular/poy.

Software User Aids

Each of the above described programs comes with a user manual and/or online help. Additional help with the use of NONA, WINCLADA, and sample data matrices can be found in a Software Tutorial written by Randall Schuh for use with *Biological Systematics*. (Available at: www.heteroptera.com)